Guide to Publishing a Scientific Paper

Guide to Publishing a Scientific Paper provides researchers in every field of the biological, physical, and medical sciences with all the information necessary to prepare, submit for publication, and revise a scientific paper.

The book includes details of every step in the process that is required for the publication of a scientific paper, for example,

- use of correct style and language
- choice of journal, use of the correct format, and adherence to journal guidelines
- submission of the manuscript in the appropriate format and with the appropriate cover letter and other materials
- the format for responses to reviewers' comments and resubmission of a revised manuscript.

The advice provided conforms to the most up-to-date specifications and even the seasoned writer will learn how procedures have changed in recent years, in particular with regard to the electronic submission of manuscripts.

Every scientist who is preparing to write a paper should read this book before embarking on the preparation of a manuscript. This useful book also includes samples of letters to the editor and responses to the editor's comments and referees' criticism. In addition, as an Appendix, the book includes succinct advice on how to prepare an application for funding.

The author has edited more than 7,500 manuscripts over the past 20 years and is, consequently, very familiar with all of the most common mistakes. Her book provides invaluable and straightforward advice on how to avoid these mistakes.

Ann M. Körner is a professional editor and writer. She has an undergraduate degree from the University of Cambridge and a doctorate in Molecular Biophysics and Biochemistry from Yale University.

Guide to Publishing a Scientific Paper

Ann M. Körner

LONDON AND NEW YORK

First published 2004 by Bioscript Press

This edition published 2008
by Routledge
2 Park Square, Milton Park, Abingdon, Oxon OX14 4RN

Simultaneously published in the USA and Canada
by Routledge
270 Madison Ave, New York, NY 10016

Routledge is an imprint of the Taylor & Francis Group, an informa business

Typeset in Garamond and Gill Sans by
Keystroke, 28 High Street, Tettenhall, Wolverhampton
Printed and bound in Great Britain by
TJ International, Padstow, Cornwall.

British Library Cataloguing in Publication Data
A catalogue record for this book is available from the British Library

Library of Congress Cataloging in Publication Data
Körner, Ann M., 1947–
Guide to Publishing a Scientific Paper/Ann M. Korner.
p. cm.
Includes bibliographical references.
ISBN 978–0–415–45265–6 (hardback) — ISBN 978–0–415–45266–3 (pbk.) — ISBN 978–0–203–93875–1 (e-book) 1. Technical writing. 2. Technical publishing. 3. Communication in science. I. Title.
T11.K68 2008
808′.0665—dc22 2007026177

ISBN 10: 0–415–45265–1 (hbk)
ISBN 10: 0–415–45266–X (pbk)
ISBN 10: 0–203–93875–5 (ebk)

ISBN 13: 978–0–415–45265–6 (hbk)
ISBN 13: 978–0–415–45266–3 (pbk)
ISBN 13: 978–0–203–93875–1 (ebk)

This book is dedicated to Sasha and Anna Hazard,
the first of a new generation of writers.

Contents

Acknowledgments

This book would never have been published without the help, over the past few years, of Shawn McLaughlin, Takeshi Seno, the staff of Yodosha Press in Japan, and Adam Sendroff. I am grateful to them all for their help and encouragement. I am also grateful to Harriet Stewart-Jones for the skill with which she prepared my manuscript for publication.

The ten most common mistakes

1. Failure to follow instructions.
2. Too many fonts on the title page.
3. Inconsistent formatting in the body of the manuscript.
4. Errors in the punctuation of references in the List of References.
5. Failure to indicate the variability and/or reproducibility of results.
6. Results that are given to a degree of accuracy that far exceeds the accuracy of measurements.
7. Graphs and histograms without indications of standard deviations.
8. Failure to distinguish between molecular mass (in kilodaltons) and molecular weight (a relative value, without units).
9. Indiscriminate use of nouns as adjectives.
10. The use of "briefly" instead of "in brief."

Introduction

There is some truth to the maxim "publish or perish." Researchers in the academic world are inevitably judged by the number and quality of their published papers; they are rarely judged by their dexterity in the laboratory, their teaching skills, or their erudition. Moreover, even the most extraordinary experimental results are of little benefit if they fail to reach the appropriate audience. Thus, the preparation and subsequent publication of a scientific paper are as important as the experiments that the paper describes. Without a published account, the value of any results is very limited. However, novices and experienced researchers often approach the writing of papers with considerable apprehension because the task is so different from work in the laboratory and yet so much depends on successful publication in an appropriate journal.

Some of the skills required for research are also required for writing a paper, for example, careful planning, organization, and attention to detail, but there is no question that writing a paper requires many skills that are quite different from those required in the laboratory. The difference is magnified when the scientist's native language is not the same as the language in which he or she has to write up the results. Thus, a young Japanese postdoctoral fellow who has mastered a complex subject, such as invertebrate evolution or mass spectrometry, and who has produced some interesting results, might find herself in the unenviable position of having to write a coherent narrative, in a foreign language, that conforms to the myriad requirements of a journal that is published on the other side of the world. A daunting task, indeed! By contrast, a young American graduate student might be unused to the discipline that is required to write a paper in a tightly defined format and might also have had little experience in writing essays or more than a few sentences at a time. Nonetheless, while the difficulties faced by aspiring authors of scientific papers vary, the goal is always the same: a paper that will find favor in the eyes of the editors and reviewers of a particular journal.

Scientists do not publish exclusively in English but English has become their common language. Thus, irrespective of whether a high-caliber journal is published in the United Kingdom, the United States, Switzerland, or Japan, it is likely that the papers in the journal will be in English. The purpose of this book is to help scientists of all nationalities to write papers that will be readily accepted for publication. I have been editing scientific manuscripts since 1985 and have edited more than 7,500 manuscripts. A quick calculation shows that, on average, I have edited a manuscript every single day for more than 20 years. This book is based on my daily experiences as an editor and it differs from similar books insofar as it focuses on those problems that I have encountered most frequently and deals in less detail with those issues that most authors tend to address correctly.

The researchers who send me their papers to correct do so before they submit them to a journal for review. In some cases, English is their native language but in many cases it is not. However, even scientists whose native language is not English have had to read many papers in English during their training and they invariably have a very good idea of what a scientific paper should look like, even if their English needs a little help. Moreover, no matter whether a scientist's native language is English, Spanish, Russian, or Japanese and no matter whether the author works at a prestigious institution or a small training college, he or she always makes some of the same mistakes. These are the very mistakes that this guide should help you to avoid.

If you have never written a paper before, you may feel overwhelmed by the task of converting the raw data in your notebooks into a coherent narrative that conforms to all the requirements of the journal of your choice. However, if you pay careful attention to all the advice in this book, you should find it relatively easy to prepare a manuscript that you can submit with confidence to your chosen journal. This book will also lead you through the prepublication process, which includes writing a letter to the editor of your target journal, responding to reviewers' comments, and resubmitting a revised version of your original paper.

Scientists send me their work to edit because they realize that, irrespective of the quality of their research, their manuscripts will always have a better reception if the text and illustrations are properly presented. In many cases, the manuscripts that I receive are accompanied by the draft of a letter to the editor, in which the author requests that his or her manuscript be considered for publication. Authors who send me their letters to the editor know that a letter that is concise and free of excessive or irrelevant information makes a good impression and increases the chances that the manuscript will pass the first important test: will the editor reject the manuscript out of hand or will he (or she) send it out for formal review? A discussion of letters to editors and examples of such letters are included in this book.

The contents of this book are organized in the same way as you should organize your thoughts and your manuscript. The chapters lead you step by step along the pathway from your decision to publish your research to the final acceptance of your paper. If you follow all the instructions in this book, you should be able to avoid all of the most common mistakes that scientists make when they write up their research for publication. Many of the examples in this book are taken from the biological sciences because that is my area of expertise. Nevertheless, the points made in each section are applicable to the publication of papers in all scientific fields. Few manuscripts are accepted without revision and, therefore, this book also includes information on how to respond appropriately to an editor's request for revisions and what to do if, in spite of your best efforts, your manuscript is rejected. Finally, since the preparation of manuscripts and applications for funding have much in common, the book ends with a brief Appendix that deals with writing a grant application.

Personal pronouns present as much of a problem when one is writing a guide as they do when one is writing a manuscript. In general, even if a researcher has worked alone, a sentence that starts, "We performed an experiment to determine whether . . ." is better than one that starts, "I performed . . ." The use of "we" is justifiable since few researchers work in total isolation. However, in this book, I shall refer to myself, your advisor, in the first person singular since this book represents my very personal approach to the art of preparing a scientific manuscript. There remains the problem of the genders of researchers and editors, who are, of course, both male and female. To avoid endless repetition of "the researcher" and "the editor" and to avoid the relentlessly annoying but politically correct "he or she" or the even uglier "(s)he," I shall consider most researchers to be male and most editors to be female. I hope that women who do research and men who edit journals will accept this approach as both economical and evenhanded.

This book, which is a distillation of all that I have learned by correcting the mistakes of others, undoubtedly contains some mistakes. I apologize in advance for these mistakes but I know that readers will find it particularly satisfying to catch me in errors that serve to demonstrate irrefutably that I too am a mere mortal.

Chapter 1

The publication of scientific papers

1.1 Why publish?

You may want to publish your research as part of your quest for fame and fortune or, at the very least, for tenure, but the only truly appropriate reason for publishing your research is to tell others about it. The purpose of your paper is to explain why you did a piece of work, how you did it, what you found, and what your findings might mean. The explanation of your methodology should be sufficiently detailed to allow the scientists in your field to repeat your work exactly, if they so choose. Your work is the foundation on which other researchers will base their future work and, as you must surely recognize, your work is based on the earlier research of others.

The publication of your work does allow you to lay claim to a particular discovery, which might be major or minor but is not, I hope, trivial, so that others will refer to your work and your contribution to the field as they continue to make progress in that field. Your contribution to the field might, in turn, bring you a modicum of fame and fortune but it is more likely that it will bring you a little closer to tenure or a promotion.

Many young scientists are under the illusion that the more papers they publish, the more they will impress the world in general and their senior colleagues in particular. A physician said to me once, "The content of my papers doesn't matter. When I'm up for promotion, it's only the number of papers that will count!" He was wrong, of course, because nobody who is in a position to make a decision about a scientist's future is just going to count that person's publications. If the people who are to decide on your next position or promotion are not experts in your field, you can be sure that they will ask several scientists in your field to comment on the significance and quality of your publications and, very probably, to rank you among your peers. Before you proceed any further, give some careful thought to the possibility that the work that you want to publish might not be as complete as it could be. If you plan to do a lengthy series of experiments over the course of a year or so, it might be better to wait until all the experiments are complete and then to write a major paper. Such a paper in a prestigious journal will count for far more than many short papers in journals that accept relatively brief communications. By contrast, if you have made a very interesting and unexpected discovery or developed a new method that will be of significant interest and assistance to your colleagues in the field, it might be better to publish a short paper or a "Letter to the Editor" right away. Now is the time to ask yourself whether you should postpone writing a paper and do some more experiments. The paper that you might write after such a delay might include an impressive amount of new information and some valuable new conclusions. It might be much better than a series of shorter papers that describe each small step along the pathway towards your final goal.

1.2 What should be published?

The only scientific research that should be published is research that is absolutely reproducible. Scientific "truths" are hard to come by and they tend to change over time. Reproducible results are the next best thing to scientific truths. The interpretation of results can mutate but, if results are reproducible, they can withstand changes in interpretation and remain both useful and relevant. Thus, before you consider publishing the results of your experiments, you need to be sure that you can reproduce them in your laboratory either exactly or, at least, within the limits of statistically acceptable fluctuations.

I am not going to discuss statistics in this book. There are many fine books about the statistical analysis of experimental results and in all likelihood you are familiar with the methods that are used in your field. However, you should bear in mind that reviewers of your manuscript will be looking carefully both to see how often you repeated your experiments and to determine how your results fluctuated when you did so. If the reviewers are not satisfied that your experiments are reproducible, they will not look kindly upon the conclusions that you draw from your results. Furthermore, if your arguments are based on differences between individual results and if these differences themselves are not statistically significant, you will also have a problem when your manuscript is reviewed. Do not start to write a paper until you are sure you can satisfy the reviewers in this regard. I once returned a manuscript in veterinary science to its author with the comment, "This manuscript is unpublishable. In your experiments, you used samples taken from only one single horse. When you have repeated your experiments with samples from several horses and shown that your results are reproducible, I shall be happy to edit a revised version of your manuscript."

Before you start writing your paper, you also should consider how many people are likely to find your work interesting. If you are working in a very small field, it is likely that your colleagues and competitors all publish in the same journals and that these journals have a relatively small circulation. Consider whether it might be better to do some more experiments to produce a piece of work that might be of greater general interest and might, thus, be publishable in a journal with wider circulation.

1.3 Who should publish?

Most research is a collaborative effort by members of a team. Such a team might include a faculty member and a small number of post-doctoral fellows, graduate students, and technicians or it might include a supervisor plus junior and senior technicians. The names that eventually find themselves on the title page of a manuscript are those of all the members of the team who contributed to the research. The final responsibility for the manuscript generally rests with the most senior member of the team, who will approve the manuscript in its final form and submit it to a journal. Junior members of any team should not attempt to publish their results without the agreement and support of the most senior member. The person with the authority to publish results is the person who "owns" the results and that person can, in general, be recognized most easily as the person who was responsible for obtaining the funds that supported the research.

Supervisors and faculty members—let us refer to them collectively as advisors—understand that the junior members of any team need to learn how to write papers if they are to advance in their careers. However, the extent of the help provided to junior members of any team varies and only rarely does an advisor make any effort to provide his junior collaborators with specific training in writing papers. A junior scientist's excitement, when he has produced publishable results for the first time in his career, is often replaced by apprehension when his advisor reacts by saying, "That's great. Now write it up!" The advisor will then await the initial draft of the paper and, depending on how patient he is, he will either work closely with his junior collaborator to revise the draft, explaining all the changes that he is making, or he will throw up his hands, disappear into his office, and prepare a new manuscript in which his junior collaborator's draft is barely recognizable.

When you "own" the research, either because you are the leader of the team and you obtained the necessary funding or because the "owner" of the research project agrees that the ideas and execution were yours alone, the responsibility for publishing your results is yours. It is your job to prepare the manuscript in its final form and to shepherd it through the publication process. It is also your job to make sure that everyone who has contributed to the research is properly recognized.

Before you start writing, it is a good idea to determine which members of your team will be co-authors and which will simply receive an acknowledgment at the end of the paper. The issue of authorship can cause serious conflicts and test friendships. It has even destroyed careers as, for example, when a department head at a major medical school insisted on having his name on a paper and then, when problems emerged about the research, it became apparent that he had not even read the manuscript.

Under optimum conditions, there should be no doubt as to who deserves to be listed as an author on a paper. The authors are those who contributed intellectually to the substance of the paper and/or who performed the experiments that are to be described in the paper. Nonetheless, even under such conditions, there can be serious squabbling about the order in which authors are listed on the title page of the manuscript. Sometimes this problem can be avoided by inclusion of a footnote that states, "The first two authors contributed equally to this work." Nonetheless, two ambitious postdoctoral fellows, Dr. Smart and Dr. Brainy, might still quarrel with their advisor because they know that their paper will be cited by others as "Smart *et al.*" or "Brainy *et al.*" and not as "Smart, Brainy *et al.*" or "Brainy, Smart *et al.*"

If the head of your section, your department, or even your laboratory insists that her name be included in the list of authors, even though she contributed nothing to the research and, in addition, she dozed through the seminars and group meetings at which it was discussed, you should take the problem to her superior for resolution. Even though the senior scientist might "own" the research, she does not have the right to be listed as an author if she made no intellectual contribution to the research. All the authors whose names appear under the title of a paper must have participated in the research described in the paper and should be able to discuss it fully, to field questions about it, and to take public responsibility for it. If the head of your laboratory or section has not made any contribution to the work described in the paper, she should not be included as an author. You can and, indeed, you should mention her name in the Acknowledgments at the end of the paper but you are not obligated to include her as an author.

The order of authors should reflect the contribution of each author, with the name of the person who contributed the most, in terms of effort and ideas, coming first. If the head of the laboratory has supervised the research, he is considered the senior author and his name is usually listed last. Before you start writing your paper, make sure that all issues relating to authorship have been resolved. If you, as the senior author, find yourself in an intractable situation, with co-authors jockeying for position, you have two choices. You can say, "I'm the boss! My decision is final," or you can say, "Since you cannot agree among yourselves, I shall write several short papers, which I shall send to mediocre journals, and you'll each get your name on a mediocre paper." The thought of their work being buried in a second- or third-rate journal should be enough to persuade the quarrelsome members of your team to settle their differences and reach a compromise.

1.4 Where should you publish?

1.4.1 General considerations

You should choose the journal to which you are going to submit your paper before you start to write it. Every journal has a different format, and every journal describes its individual formatting requirements in a section entitled "Instructions to Authors." If you have chosen a target journal before you start writing, you can follow the specific instructions for contributions to that journal as you prepare your manuscript for submission.

Since your work follows and extends similar work in your field, you probably already know where research such as yours is published.

You have used methods described in previously published papers and your working hypothesis is based on the conclusions that others have published. The papers to which you will refer in your paper were probably published in a relatively small number of journals and you should choose from among them, in particular, if you work in a very circumscribed field, such as, for example, clinical biomechanics or astroparticle physics. However, if your work is of high caliber and broader interest, you might try to publish in *Science*, *Nature*, or the *Proceedings of the National Academy of Sciences of the United States of America* (known to most scientists as "*PNAS*"), all three of which are read by scientists in a wide variety of disciplines. You should bear in mind, however, that the wider the readership of a particular journal, the harder it is to publish in that journal. For example, the editors of *Science* accept only about 10 percent of the papers submitted for review and they reject approximately 65 percent of all manuscripts that are submitted to *Science* within a mere week to ten days of receiving them. The figures for *Nature* are similar. In 2005, the editors of *Nature* received a total of 8,943 manuscripts, of which they published only 915, and they returned most of the papers that they rejected to the authors without review.

There are thousands and thousands of journals. The website of the Mulford Library of the Medical College of Ohio, http://mulford.meduohio.edu/instr/, provides links to the Instructions to Authors of 3,500 journals and those are only the journals that deal with the biological and medical sciences. There are also large numbers of journals that serve researchers in the physical sciences and a useful link to the Instructions to Authors of many of them is http://www-library.lbl.gov/library/public/tmLib/journals/LibJourInstr.htm. A similar site with links to chemistry journals is http://www.ch.cam.ac.uk/c2k/cj/ and a site with links specifically to journals in physics is http://info.ifpan.edu.pl/journal.html. Geologists can find a list of 3,000 journals, published worldwide, on the website of the American Geological Institute (http://www.agiweb.org/georef/about/serials.html), as well as a link to an abbreviated list of the 99 journals that the Institute considers to be "priority journals."

You can, if you like, take comfort in the fact that, with so many journals published annually, you are bound to get your paper published somewhere. To some extent, you are right. The companies that publish journals want to make money. They make money by selling subscriptions to libraries and educational institutions. If they are to fill a certain number of issues every year, they need papers from people like you. However, in the first instance, you should aim high and try to publish your paper in the best possible journal in your field.

1.4.2 Specific considerations

There are two practical matters that you need to address in your choice of journal. Many journals now require electronic submission of manuscripts and the instructions for such submissions can be quite complicated, in particular, when Figures and Tables are part of the manuscript. Before you make the final choice of your journal, look carefully at the most recent issues and establish to your own satisfaction that your research is appropriate for the journal. Then study the Instructions to Authors carefully to make sure that you can submit your manuscript in the required format. Your study of the Instructions to Authors might lead you to a discovery that surprises you. The publishers of some journals make you pay for the privilege of having your manuscript appear in one of their journals! In 2007, the charges for publication in one particular journal were as follows:

- Electronic manuscript: $105 per page
- Paper manuscript: $150 per page
- Color Figure surcharge (for the print edition): $100 per page.

If you think these prices are astronomical, you are right. They were the prices for publishing in the *Astronomical Journal*! It pays, in this regard, to study butterflies instead of the stars. The *Journal of the Lepidopterists' Society* charged only $50 per page and asked a mere $25 from those lepidopterists who are not associated with an academic institution. When page charges were introduced, there was much grumbling. Nevertheless, the practice is now widespread and generally accepted. You need to make sure that you have the funds to pay for publication of your work and also for the reprints (known in Britain as offprints) that you will want to send to your colleagues. The payment of page charges has led to a peculiarly amusing anomaly: papers that appear in journals with page charges are referred to legally as "advertisements."

1.5 Manuscripts for biomedical journals

If you are planning to write a paper for a biomedical journal, it would be a good idea at this point to read the "Uniform Requirements for

Manuscripts Submitted to Biomedical Journals" of the International Committee of Medical Journal Editors, which you can find on the internet at http://www.icmje.org/index.html. At this address on the internet, you will also find a link (http://www.icmje.org/sponsor.htm) to an article on "Sponsorship, Authorship, and Accountability" by the same committee, which you should read carefully if your research is sponsored by a "for profit" or commercial organization rather than a non-profit or government organization. You will find on this website a careful analysis of problems related to conflicts of interest and, if you have any doubts as to possible conflicts of interest between you and your research on the one hand and your source of funding on the other, you should address them now, before you proceed any further.

Chapter 2

Before you start writing

2.1 Instructions to Authors

Every journal that is published supplies specific Instructions to Authors. If every author followed these instructions when preparing a manuscript, there would be only a limited need for a book such as this one. Every journal has its own specific set of instructions and, even though the instructions for submissions to all the journals produced by a particular scientific publisher tend to be very similar, different publishers provide different instructions. Authors tend to get lazy or, perhaps, I should say that they get carried away with the idea of publishing their research and, as a result, they prepare their manuscripts according to some preconceived idea or on the basis of a manuscript that they have published previously. Careful preparation before starting to write will save you time and spare you frustration later on. In publishing, as in everything else in life, it pays to READ THE INSTRUCTIONS.

When you have chosen the journal in which you hope to publish your work, spend some time looking at the papers in several of the most recent issues to get an idea of the style and format of papers in the journal. No matter how original your research might be, your goal now is to produce a manuscript that is identical, in terms of style and format, to papers that have already been published in your target journal. Your review of previously published papers in your target journal should also provide confirmation that you have chosen the appropriate platform for your paper. Some journals include their Instructions to Authors in only one issue per year and some include these instructions in every issue. Many journals now refer prospective authors to instructions that are published on the internet. Irrespective of whether you find the instructions in a printed copy of the journal or on the internet, you need your own hard copy of these instructions, in other words, a copy on paper, so that the instructions are available for instantaneous reference at every stage in the preparation of your manuscript. Thus, you should photocopy printed instructions or download and print out the instructions that are provided on the journal's website.

As mentioned in Chapter 1, the website of the Mulford Library of the Medical College of Ohio, http://mulford.meduohio.edu/instr/, provides alphabetized links to the Instructions to Authors of 3,500 journals in the biological and medical sciences. A similarly comprehensive list of chemistry journals can be found at http://www.ch.cam.ac.uk/c2k/cj/, and a list of physics journals can be found at http://info.ifpan.edu.pl/journal.html.

Once you have generated your own personal copy of the Instructions to Authors of your target journal, read the instructions from start to finish in order to get a general impression of what is required. Then take a highlighter pen or an old-fashioned red pen and highlight or underline each of the very specific instructions that define the presentation of your text. Finally, reread the instructions, focusing on the parts that you have underlined. This close attention to the instructions might seem to be a waste of time but you would be surprised at the extent to which the publication of papers has been delayed by failure to attend to the details of presentation. Moreover, as I shall continue to emphasize, a manuscript that is properly formatted receives a better reception than one that does not conform to the requirements set out in the Instructions to Authors.

When you have studied and absorbed the critical points in the Instructions to Authors of the journal to which you will submit your manuscript, you are almost ready to start writing. However, before you start, I would like you to read the next section, which focuses on common grammatical errors. Then, armed with a highlighted copy of the Instructions to Authors and with tools that will help you to avoid the most common grammatical mistakes, you will be ready, finally, to start preparing your manuscript.

2.2 Common grammatical mistakes

2.2.1 Why does grammar matter?

The dominant purpose of publishing your work is, as I have noted, to share your results and to provide information about your methodology that is sufficiently detailed to allow others to repeat your experiments. The descriptions of your methods and results must, therefore, be absolutely free of ambiguity, and the possibility of misinterpretation must be minimal. Correctly formulated sentences, which conform to the rules of grammar, are remarkably effective tools for unambiguous communication. The scientific community is spread out all over the world and the common language of scientists is English. If your manuscript is written clearly and correctly, your colleagues everywhere will be able to understand it. If your English is ungrammatical, colloquial, or sloppy, your colleagues will have a much harder time figuring out what you are trying to say. This is not a book about grammar, just as it is not a book about statistics, but there are a few grammatical mistakes that crop up so frequently that it is worthwhile summarizing them here in the hope that you will remember to avoid them when you write your paper.

2.2.2 Spelling and consistency

Some words have British and American versions, for example, "analyse" and "analyze," "sulphate" and "sulfate." If you are submitting your paper to a British, Canadian, or European journal, you should try to use British spelling exclusively. If you are used to American spelling, you should check the Instructions to Authors of your target journal to see whether American spelling is also acceptable. However, no matter whether you use British or American spelling, consistency is the key. If you use the British spelling of some words, you must use the British spelling of all words. If you are submitting your paper to an American journal, you should use American spelling consistently. In some cases, two alternative spellings of a given word are acceptable, for example, "labeled" and "labelled," "focused" and "focussed." The choice is yours in each case but, again, you must be consistent and use one version or the other throughout your paper. Indeed, throughout your paper, you should strive for consistency in spelling, abbreviations, and units.

You should pay particular attention to these issues if your colleagues have supplied you with drafts of the various sections of your paper that deal specifically with experiments that they have performed. A paper should look and read as if it was written by a single person and not by a committee.

2.2.3 *The active versus the passive voice*

It is always best to write simple, declarative sentences. In other words, you should try to avoid the passive voice as far as possible. For example, it is better to say, "We studied the behavior of orangutans in the wild," than to say, "A study was performed of orangutans in the wild." Similarly, it is better to say, "We synthesized several borane complexes and studied their structures" than to say, "The synthesis and structures of several borane complexes are described." Moreover, many authors forget that certain nouns are derived from verbs, for example, "preparation" is derived from "prepare." If you bear this relationship in mind, you are less likely to write, for example, "A preparation of DNA was made by . . ." and more likely to write, "DNA was prepared by . . ." or, avoiding the passive voice altogether, you should write, "We prepared DNA by . . ." Similarly, you should avoid writing, for example, "Separation of the products of the reaction was performed by reverse-phase chromatography"; write "We separated the products of the reaction by reverse-phase chromatography" instead.

2.2.4 *The incorrectly related participle*

A participle (an "-ing" word, such as standing, running, and jumping) always relates to the grammatical subject of the sentence (the person or thing that governs the main verb). For example, the sentence, "The man walked down the street wearing a blue hat," is correct and is easily understood to mean that a man with a blue hat was walking down the street (the man is the grammatical subject). Compare this sentence with the sentence, "The man walked down the street leading to the center of town." There is a problem here because, without question, it was the street that led to the center of town and not the man. If the man, the grammatical subject of the sentence, were doing the leading, we would expect something along the lines of "The man walked down the street, leading the circus to the center of town." To avoid ambiguity, in the previous example, we have to write, "The man walked down the street that led to the center of town."

In a scientific paper, use of an incorrectly related participle results in sentences such as, "The cells were grown in ABC medium containing glycine." The correct version is, "The cells were grown in ABC medium, which contained glycine." Similarly, the following sentence is incorrect since the subject of the sentence is "the cells" and the participle does not relate to them: "The cells were shown to contain pigmented granules using the electron microscope." In spite of significant progress in genetic engineering, pigmented granules are not yet able to use an electron microscope. The sentence should read, "Using the electron microscope, we observed that the cells contained pigmented granules." To avoid mistakes with participles, always ask yourself whether the participle is correctly related to the grammatical subject of your sentence.

2.2.5 The use of "that" and "which"

Compare the following sentences, "The cells grew in the enriched medium that contained calcium ions" and "The cells grew in the enriched medium, which contained calcium ions." The first sentence emphasizes the observation that the cells grew in an enriched medium that contained calcium ions, as distinct from some other medium, for example, an enriched medium without calcium ions. The second sentence emphasizes that the cells were able to grow in the enriched medium, and the sentence includes, at the same time, the information that the medium contained calcium ions. By asking yourself whether you can insert an imaginary "incidentally" into your sentence, you can often determine whether you need a "that" or a "which." If you can insert "incidentally" without altering the meaning of the sentence, you should write "which" and not "that." Thus, when you apply this criterion to the two examples given above, you have to ask yourself whether the presence of calcium ions was important or incidental. If it was important, you need to write, "The cells grew in the enriched medium that contained calcium ions." If the presence of calcium ions was incidental, you need to write, "The cells grew in the enriched medium, which contained calcium ions." Remember that if you use "which" rather than "that," the word "which" has to be preceded by a comma.

2.2.6 Nouns as adjectives and the problems that they cause

Consider a phrase that has become accepted in endocrinology (the study of hormones and their actions), "the growth hormone receptor antagonist." Now consider the phrase, "the cloned growth hormone receptor antagonist." The first phrase, which refers to a molecule that binds to the receptor for growth hormone, is not ambiguous and is widely used by endocrinologists who study growth hormone. The second phrase presents some problems: does the adjective "cloned" refer to "growth hormone," to "growth hormone receptor," or to "antagonist?" It is too late to complain that the cumbersome term "growth hormone receptor antagonist" has become firmly entrenched in the jargon of endocrinologists but it is necessary for authors to take extra care when modifying terms in which nouns are used as adjectives. The unambiguous use of the adjective "cloned" leads to the following possibilities: "the antagonist directed against the cloned receptor for growth hormone;" "the cloned antagonist directed against the growth hormone receptor;" and "the antagonist directed against the receptor for cloned growth hormone." Each of these possibilities is plausible and, therefore, the author must choose the one that is appropriate to avoid all ambiguity. It is particularly important to avoid ambiguity if you anticipate an international readership for your paper.

The following phrase, from an actual manuscript that I was editing while writing this chapter, provides another example of what happens when an author attempts to compress a complicated concept into a phrase in which a noun is modified by a string of nouns and adjectives: "mouse marrow-derived macrophage colony-stimulating factor- (M-CSF-) dependent monocytes." In this phrase, the noun that is being described is "monocytes." These monocytes were derived from mouse marrow and their growth was dependent on macrophage colony-stimulating factor (M-CSF). The corrected version of this phrase is, therefore, "monocytes, derived from mouse marrow, whose growth was dependent on macrophage colony-stimulating factor (M-CSF)." Scientists who work in the same field as the author of the paper in which the original phrase appeared might say that they had no trouble understanding what the author meant. However, anyone from a slightly different field might not have found it so easy to figure out what the author was trying to say.

Let us consider, as another example, the title of a very specialized paper in chemical physics, "Restricted density-functional linear response theory calculations of electronic g-tensors." I do not doubt that scientists who make such calculations understand this title perfectly but, in this convoluted title, the word "calculations" is modified by three adjectives, namely, "restricted," "density-functional," and "linear," one of which itself includes a noun, namely, "density," as well as two nouns, namely, "response" and "theory." Such a combination of nouns and adjectives, which appear to modify one another willy-nilly, is totally opaque to the non-specialist and demonstrates how scientific jargon threatens to divide scientists into more and more tiny subgroups, none of which speaks the language of any other subgroup. Absolute clarity allows readers whose native language is not English to read scientific papers with ease and without misunderstandings. It also helps scientists in related disciplines to understand each other's papers.

2.2.7 "This" is often incorrect

The word "this" is an adjective. When "this" stands alone, it is likely that the author has replaced an opportunity for clarity by opacity and, in order to avoid ambiguity, he needs to find the noun or a noun to which "this" refers and set it down after "this." This is an important point. See what I mean? What exactly is the "important point" that I am making here? Is it the fact that the author has replaced an opportunity for clarity by opacity; is it the fact that it is necessary to avoid ambiguity; or is it the fact that the author needs to find the noun to which "this" refers and set it down after "this?" My sentence, "This is an important point," should read, for example, "These issues are important." Other possibilities are "Thus, it is important to remember that 'this' is an adjective and must modify a noun" and "The writer's goal should be clarity and not opacity."

In scientific manuscripts, "this" is often used incorrectly. For example, after a lengthy description of his results, a lazy author might start a new sentence with "This showed . . ." without specifying which of the preceding observations "showed" something. Consider the sentence, "The cells divided in the modified medium and formed clumps that were visible to the naked eye." Four observations are included in this sentence, namely, the cells divided, they divided in modified medium, they formed clumps, and the clumps were visible to the naked eye. Thus, it makes no sense to follow the sentence in question with a sentence that starts, "This showed . . ." It is necessary to start the sentence by referring to one or more of the specific observations, as follows, "The formation of clumps showed . . ." or "The division of cells in the modified medium showed . . ." or "The formation of large clumps of cells that could be seen with the naked eye showed . . ." If you always remember that "this" is an adjective and cannot stand alone, you will avoid this mistake.

2.2.8 The incorrect use of "due to"

The phrase "due to" can only link two nouns. It cannot be used as an adverb. Thus, it is correct to say, "The formation of blue colonies was due to the presence of an enzyme in the cells." "Formation" (noun A) was due to "presence" (noun B). It is incorrect to say "Due to the presence of an enzyme, we saw blue colonies" (here, "we" and "due to the presence of an enzyme" are not related directly). It is also incorrect to say, "Blue colonies were formed due to an enzyme in the cells;" the correct version is, "Blue colonies were formed as a result of the activity of an enzyme in the cells." When you use "due to," make sure that a specific "noun A" is due to a specific "noun B." You should make an effort to avoid this mistake. Careless writing is often due to laziness.

2.2.9 "Types," "kinds," and "classes"

Writers of all types, and not only scientists, often make mistakes when they write about "types," "kinds," and "classes." The easiest way to explain the correct usage is by example.

Consider the following phrases: one type of child; one kind of parent; and one class of children. The respective plurals of "type," "kind," and "class" are "types," "kinds," and "classes." Thus, if there are, for example, two of each, we form the plurals as "two types of child," "two kinds of parent," and "two classes of children." Similarly, when writing about cells or subatomic particles, you should recall the following examples: "two types of cell," "two kinds of cell," and "two classes of cells"; and "many types of subatomic particle," "many kinds of subatomic particle," and "many classes of subatomic particles." The expressions "two types of cells," "two kinds of cells," "many types of subatomic particles" and "many kinds of subatomic particles" are all incorrect. You should strive to avoid this type of blunder, this kind of error, and these classes of mistakes.

2.2.10 "None" means "not one" and is singular

You should think of the word "none" as being an abbreviated form of "not one." If you do so, you will avoid the common mistake of using "none" as if it were a plural noun. It is incorrect to say, "None of the results reported by Smart *et al.* are in agreement with those reported by Brainy *et al.*" The correct version is, "None of the results reported by Smart *et al.* is in agreement with those reported by Brainy *et al.*" If, as you are writing, you remember that "none" really means "not one" and "none of" really means "not one of," you will avoid a very common mistake.

2.2.11 Some common problems with hyphenation

An excellent discussion of correct hyphenation can be found in the invaluable writers' aid, *The Chicago Manual of Style*, which is published by the University of Chicago Press. If you are serious about writing, I strongly recommend that you buy this book. In it, you will find all the rules that apply to hyphenation in English grammar. You will learn from this invaluable text that compound adjectives that include well, ill, better, best, little, and lesser are hyphenated before the noun unless the adjective is further modified by, for example, "very." This rule results in the following correct phrases: "well-known theorem," "very well known theorem," and "the theorem is well known." Another useful rule is that adjectival compounds with "-fold" are spelled as a single word unless they are formed with figures. This rule yields phrases such as "a tenfold increase" and "a 35-fold increase." There is one more rule that is worth repeating here, namely, that when a prefix stands alone, it must be followed by a hyphen, for example, "exo- and endothermic reactions." Similarly, there is a hyphen after the first number in phrases such as "four- to five-day-old cultures" and "two- or three-step reactions."

2.2.12 Hyphenation and abbreviations

Abbreviations need to be written out in full when they are first mentioned, but you should take care if that first mention happens to be in the middle of a hyphenated phrase. Papers in cell biology often include constructions such as, "The protein was visualized with fluorescein isothiocyanate (FITC)-conjugated antibodies." Here it is incorrect to introduce the abbreviation for fluorescein isothiocyanate, FITC, in the middle of this phrase. The correct version is "The protein was visualized with fluorescein isothiocyanate-conjugated (FITC-conjugated) antibodies." You should only introduce the abbreviation for a word that is part of a hyphenated phrase at the end of the phrase, at which point you should use the abbreviation in a repeat of the entire hyphenated phrase.

2.2.13 Numbers and hyphens

It is standard practice to write out numbers from one to ten as words (for example, write "one" for "1") unless the number is followed by abbreviated units. For example, you should not write "two g/l" for "two grams per liter." The correct version is "2 g/l."

You should not start a sentence with a figure, such as 1, 45, or 100. To avoid this mistake, you should use a word to lead into the sentence. However, if you write the number out in full, you should also write the units out in full, which becomes rather cumbersome. For example, instead of writing, "1 ml of the solution was mixed with . . .," you would have to write, "One milliliter of the solution was mixed with . . ." To avoid this problem, you could write, "An aliquot of 1 ml of the solution was mixed with . . ." or, avoiding the passive voice altogether, you might write, "We mixed 1 ml of the solution with . . ."

Authors often make mistakes when numbers are used in hyphenated phrases, such as "two-year-old" and "12-fold." The common mistakes are of two types. In one type, the first of two necessary hyphens is missing. Consider the phrase "two year-old horses," which means "two horses that are both one year old." Compare this first phrase with the phrase "two-year-old horses." This latter phrase refers to some undefined number of horses that are two years old. You need to be sure to use the correct number of hyphens. If you are using numbers greater than ten, you should use figures, rather than the words, as, for example, in the phrase "14-year-old children." If you are, in fact, referring to 14 children, each of whom is one year old, you should say, "14 one-year-old children" rather than "14 year-old children."

The second type of common mistake occurs frequently when an author refers to the range covered by a certain parameter, for example, age. The following is an example of correct usage: "We studied a group of 11- to 15-year-old children." Be sure to remember the hyphen after the first number when you use phrases such as this one (for example, "650- to 700-ml aliquots," "two- to three-year period" and "200- to 300-μm difference"). Similarly, you need to remember the hyphen after the first number when you use "-fold," as in the examples, "We observed a three- to sixfold increase in the rate of . . ." and "There was a 200- to 300-fold increase in the number of . . ."

2.2.14 Lists and semicolons

The semicolon is a useful punctuation mark, especially when you are making a list of complicated items. For example, if you are making a cake, you need the following ingredients: butter, preferably unsalted but salted can be used if unsalted butter is not available; sugar, either granulated or powdered; eggs, which should be as fresh as possible; and flour, either plain or self-raising. Without semicolons, this list would read as follows: butter, preferably unsalted but salted can be used if unsalted butter is not available, sugar, either granulated or powdered, eggs, which should be as fresh as possible, and flour, either plain or self-raising. As you can see, the semicolons allow each ingredient and its description to be separated from the others without confusion or ambiguity. Similarly, in discussing three sets of primers for the polymerase chain reaction (PCR), it is necessary to use semicolons to separate the various pairs of primers, as demonstrated in the following sentence. "The primers for PCR were forward 1, ATTGCCATCCAG, and reverse 1, CGGATTAACGCC; forward 2, ACCGTTGCAAGT, and reverse 2, CCAGTTGACTGA; and forward 3, ACGACTGCATGC, and reverse 3, ACCAGTTGCAGT." As you can see, each pair of primers is separated from the next one by a semicolon. Without semicolons, the same sentence would be much harder to understand: "The primers for PCR were forward 1, ATTGCCATCCAG, and reverse 1, CGGATTAACGCC, forward 2, ACCGTTGCAAGT, and reverse 2, CCAGTTGACTGA, and forward 3, ACGACTGCATGC, and reverse 3, ACCAGTTGCAGT." The semicolons separate items, namely, pairs of primers, that need to be considered separately, just like the butter, sugar, eggs and flour that go into a cake.

You will also need to use semicolons when you provide the sources of the various items that you used in your experiments. In order that other scientists can repeat your experiments exactly as you performed them, if they so choose, you must provide the sources of all materials and instrumentation that you used. Each item must be identified in sufficient detail to allow another scientist to obtain exactly the same item. The details of each item must include the full name, trade name or model number, when relevant; the name of the manufacturer; and the location of the manufacturer (did you notice the use of semicolons in this sentence?). Here are some examples of the correct format: "We examined the ultrastructure of the xylem under an electron microscope (model 1200; ElectroMic Co., Ltd., Placeville, NY, USA)." "The samples were dissolved in deuterium oxide (99.9 atom % D; Isochem Co., Ltd., La Buena, CA, USA)." "The products of transcription were blotted onto nitrocellulose filters (BD45; Sengen Chemical Co., Eikoyama, Japan)." In the first example, the model number is separated by a semicolon from the manufacturer's name and address. In the second example, details of the deuterium oxide are separated by a semicolon from the manufacturer's name and address. In the third example, the product identification number is separated by a semicolon from the manufacturer's name and address. In each of these examples, semicolons separate the items that are listed in parentheses.

The semicolon is your friend. It helps to separate the individual items in a list of items from one another and, in this way, it helps the reader to move from one item to the next without any confusion.

2.3 Reference books

The explanations, suggestions, and examples given above should help you to avoid some of the most common mistakes, in terms of grammar and spelling, that I find when I correct scientific manuscripts. A good grounding in grammar will help you to write clearly and correctly. The more you read, the better you will write. I was asked some years ago to develop a course in scientific writing for undergraduates. I turned down the invitation, suggesting instead that the institution in question should require that all its science students read the complete works of Jane Austen. I am not going to suggest that you do the same before you start to write your paper but I do recommend that you read as many literary classics as you can. Your writing and vocabulary will improve and the process will be relatively painless.

Every writer needs a good English dictionary, for example, *Webster's Ninth New Collegiate Dictionary.* The "spell-check" function of your computer's word-processing program will help you to avoid many spelling mistakes but you cannot rely on it entirely. If you are unsure of the meaning of a word or the past tense of a verb, for example, you will still find a dictionary to be very useful, even though you can find such information on the internet if you know where and how to look for it. As noted above, another invaluable resource for writers in any field is *The Chicago Manual of Style.* You will also not regret investing in reference books that relate directly to your field. For example, you might purchase *Stedman's Medical Dictionary* (published by Williams and Wilkins) and *Gray's Anatomy* (published by Gramercy Books and distributed by the Outlet Book Company, a division of Random House), if your research is even remotely medical, while *Hortus Third* (published by Macmillan) is essential if you are a botanist. In spite of the vast amount of information on the internet, you will find it useful to have classical reference works within arms' reach when you are writing your paper.

If you want to travel light, you can rely on the internet but you should bear in mind that websites can be less reliable than printed works of reference. Nonetheless, if you are unsure about the spelling of a particular word or chemical compound, you can easily search for websites that mention the word using "Google" (http://www.google.com) or some other search engine. If you have made a small spelling mistake, typing, for example, "Caenorabditis elegans" instead of the correct version in the appropriate box on the Google website, you will immediately be asked, "Did you mean *Caenorhabditis elegans*?" If there are two ways to spell a technical term, you can search for each using Google and then compare the number of results that you get, choosing the version that yields the largest number of results. You can be confident that this version is the one that is most commonly used. If you are trying to decide whether to use the spelling "tetracycline" or "tetracyclin" for a common antibiotic, Google will tell you that there are more than 4 million sites that refer to the former and only about 400,000 sites that refer to the latter, making your decision very easy.

Many writers have come to rely on the "spell-check" function that is provided with computerized word-processing programs. By all means, use this function to eliminate simple spelling mistakes but do not rely upon it to do all your work for you. If you do not read through your work carefully, strange errors may creep into it. In a manuscript that I edited some time ago, the author had typed in "fluorescent grouts", obviously meaning "fluorescent groups," because his finger hit the wrong key on his keyboard. At his command, the computer checked his spelling, recognized that "grouts" was an authentic word and failed to replace it by "groups." Thus, the text that I received included a discussion of "fluorescent grouts," which might be of interest to someone with a dark bathroom but has no place in a scientific paper. Another frequent indicator of an author's reliance on a computerized spell-check program is replacement of the word "summary" by "summery." Such mistakes make a bad impression and every effort should be made to avoid them.

Finally, you should remember that people are the most useful tools of all. You can avoid submitting a manuscript that contains obvious errors by asking a few of your colleagues, at your own or another institution, to read your manuscript before you send it to your target journal. Your colleagues can help you insure that your manuscript contains no serious mistakes. You should ask them to confirm, as far as possible, the appropriateness of your methodology, the quality of your results, and the validity of the conclusions that you have drawn from your results. Your colleagues can also help you to eliminate any trivial typographical errors before your manuscript is scrutinized by the editor and the reviewers that you hope to impress. If your colleagues make a significant contribution to your work at this stage, remember to express your gratitude to them in the Acknowledgments section of your manuscript. Remember, too, that you are obliged to reciprocate in kind when your colleagues ask you to look over drafts of their papers.

Chapter 3

The title page

The title page is your first opportunity to make a good impression. Mistakes on the title page can have a considerable negative effect on the reception of your manuscript by the editor and the reviewers that she chooses to assess the quality of your work. Nonetheless, you would be surprised how many authors make mistakes on the title pages of their manuscripts. Make sure that you are not one of them. The exact format of the title page will depend on the Instructions to Authors of the journal to which you will submit your manuscript. As you follow these instructions, you should also keep the following points in mind.

3.1 The choice of title

You need to give careful thought to the title of your manuscript. The Instructions to Authors may specify the maximum length of your title but, even if they do not, you need to keep your title relatively brief. Do not try to include every result and conclusion in your title and be sure to mention the system in which or the organism with which your study was performed.

The best titles are short, declarative sentences that describe the major conclusion suggested by the results. For example, "The enzyme responsible for the amino-terminal modification of sputase in mice is a basic cytoplasmic protein" is clear, concise, and gives all the information the reader needs. Some of the worst titles begin with the words "Studies of . . ."; for example, "Studies of the amino-terminal modification of sputase in the mouse" or "Studies of the diffusion of photons in turbid media." If your work is purely descriptive, your title should include the parameter(s) or feature(s) that you have studied and the system in which you have studied it, for example, "The biochemical and physical properties of sputase from the mouse" or "Diffusion coefficients for the movement of photons in turbid media." Your title is the only way that you have to catch the eye of prospective readers. When thumbing through a journal or scanning a "Table of Contents," nobody is going to give a second thought to a paper with a boring, confusing, or uninformative title. Boring and uninformative titles often begin with the words "Studies of . . ." and confusing titles often include a long string of nouns used as adjectives, such as, "The effects of the cloned mouse fibroblast insulin receptor antagonist combinatogen on mouse lung explants *in vitro.*" In this title, an adjective and five nouns "describe" the fictitious protein combinatogen and only an expert in the field would be able to unravel the meaning of this title.

Uninformative titles often include unusual abbreviations that the author has defined in the text but not in the title, for example, "The effects of the mouse NC134 cell IR antagonist C27 on mouse lung explants *in vitro.*" Such titles also often fail to mention the experimental system in which the author's study was performed, for example, "The effects of combinatogen *in vitro.*" A reasonable title, based on our fictitious example, would be, "Combinatogen, an antagonist of insulin receptors on murine fibroblasts, stimulates secretion of combinase from murine lung explants *in vitro.*" If combinatogen were, in fact, a well-characterized protein, the title could be abbreviated to "Combinatogen stimulates secretion of combinase from murine lung explants *in vitro.*"

Your paper might be one in a series of papers from your laboratory and you might be tempted to use a title such as "Studies of combinatogen, IV," if your paper is the fourth in the series. However, this title is extremely uninformative and will be relevant only to those scientists who have followed your publications so avidly that they are familiar with and remember the first, second, and third papers that describe your work on combinatogen. Your title should refer specifically to the major discovery in your most recent study so that the subject of your paper is clear to all potential readers. Moreover, even though your paper describes your most recent discovery and your results are, by definition, "new," you should avoid using the words "first" and "novel" in the title of your paper. As noted recently by the editorial board of the *Journal of the American Chemical Society,* these adjectives "as used in titles of manuscripts, are generally abused, overused and are not really necessary."

When I am trying to improve a manuscript with a title that is boring, confusing, or uninformative, I ask myself two questions: "What is the single most important point made in this paper? How would I tell another scientist, in one short sentence, what this paper is all about?" The answer to these questions often forms the basis for an appropriate title. If you are casting around for a good title, ask yourself the same questions. The answers might help you to compose an appropriate title for your paper.

Finally, before you move on from your title, check your target journal to see whether you need to add a footnote to your title that specifies the source(s) of funding for your research.

3.2 The running title

At the top of each page of most published papers there is a brief "running title" (or "header") that characterizes the research described on that page. The Instructions to Authors usually include the exact specifications for the running title, including the number of letters and spaces allowed—normally 50 or 60. In most cases, you must remember to count the spaces as well as the letters and you should on no account provide a running title that is longer than the journal allows. Editors and publishers will be totally inflexible in this regard so you must make the necessary effort to meet their specified criteria.

The running title can present a challenge. You have compressed your work into the one brief, declarative sentence or description that is the title of your paper. Now you have to compress this title into a running title of not more than 50 or 60 letters and spaces. It is helpful, at this point, to consider the function of the running title. When a reader holds a journal in her hands and flips through the pages, a Figure or a Table might catch her eye. The running title, printed at the top of the page, provides her with a useful reference point and, if the running title appears sufficiently interesting, she will turn to the first page of the paper and start reading it from the beginning. The main words in the running title should be identical to words in the main title. Otherwise, on finding that the title bears little relationship to the running title, the reader might feel that she has been misled and will return to skimming through the journal for something else closer to her areas of interest. If the running title contains abbreviations that are not in general use, the reader who stops momentarily to look at an illustration and then glances at the running title is less likely to turn to the beginning of the paper and start to read it than she is if the running title makes perfect sense.

The rules for writing a running title are as follows: it should not be longer than is specified in the Instructions to Authors; it should include words from the main title and should not include concepts or references to materials that are not in the main title; and it should not include unusual abbreviations. If you find that it is impossible to obey these rules, you should change the title of the paper so that you can generate a running title without breaking these rules.

3.3 The authors' names and relevant footnotes

The quickest way of insuring that you are using the appropriate format for the names of the authors of your manuscript is to refer to papers that have already been published in your target journal. Some journals require full names, some require last names and initials only. Some journals require academic degrees in addition to authors' names and, in such cases, punctuation becomes very important, with periods (known as full stops to those who speak British English), commas, and semicolons all over the place. There are numerous possible permutations and combinations of first names, initials, last names, and degrees, as shown in the following examples. Take a moment to look at these examples carefully to identify the differences between them:

Peter Brainy, Mary Gifted, and Henry Smart

P. Brainy, M. Gifted, and H. Smart

Brainy, P., Gifted, M., and Smart, H.

Peter Brainy Ph.D., Mary Gifted M.D., and Henry Smart Ph.D., M.D.

P. Brainy Ph.D., M. Gifted M.D., and H. Smart Ph.D., M.D.

Brainy, P., Ph.D.; Gifted, M., M.D.; and Smart, H., Ph.D., M.D.

Your target journal probably requires one of these specific formats or a very similar format and you must type the authors' names in exactly the right format. Your work on the names of the authors is not yet complete. If all the authors contributed equally to the work described in the paper and still work at the same institution, you will only have to provide the name of a single institution under the authors' names. It is likely, however, that some of the authors have moved to other institutions or performed the work that is described in your paper at some other institution. In such cases, all the institutions have to be included under the authors' names, with appropriate designations, as shown in the following example:

> Peter Brainy[1], Mary Gifted[2], and Henry Smart[3]
>
> [1]Present address: Institute of Scientific Research, University of Flatland, Podunk, CA, U.S.A.
>
> [2]Bioeconomica and Co., Ltd., Podunk, CA, U.S.A.
>
> [3]Dept. of Important Research, University of Erewhon, Utopia, CA, U.S.A.

At this point, you may also wish to mention that Peter Brainy and Mary Gifted played equally important roles in the study, as indicated by the superscript "4" in the example that follows:

> Peter Brainy[1,4], Mary Gifted[2,4], and Henry Smart[3]
>
> [1]Present address: Institute of Scientific Research, University of Flatland, Podunk. CA, U.S.A.
>
> [2]Biotechnology and Co., Ltd., Podunk, CA, U.S.A.
>
> [3]Dept. of Important Research, University of Erewhon, Utopia, CA, U.S.A.
>
> [4]The first two authors contributed equally to this work.

Some journals use numbers for superscripts, as in the example above. Others use symbols (for example, *, ¶, §, and †) and the order in which you should use them is generally specified in the Instructions to Authors. As mentioned above, it is often helpful to refer to the format of papers that have already been published in your target journal. You can use the published format as a template but you should then confirm that you have also adhered exactly to the Instructions to Authors.

3.4 The author for correspondence

The title page usually includes the name of the author who is designated as the "Author for all correspondence and requests for reprints" or, more simply, the "Corresponding author," or "Author for correspondence." Sometimes the author's name is accompanied by an e-mail address and fax number. For the exact wording and format, refer again to the Instructions to Authors or to published papers in your target journal. If the information relating to the author who will be responsible for all correspondence appears as a footnote, you will have to modify the list of the names of the authors as follows:

Peter Brainy[1,4], Mary Gifted[2,4], and Henry Smart[3,5]

[1]Present address: Institute of Scientific Research, University of Flatland, Podunk, CA, U.S.A.

[2]Biotechnology and Co., Ltd., Podunk, CA, U.S.A.

[3]Dept. of Important Research, University of Erewhon, Utopia, CA, U.S.A.

[4]The first two authors contributed equally to this work.

[5]Author for all correspondence.

For the exact format, for example, the use of numbers or symbols and the order of symbols, you should, as I have mentioned, refer to your target journal. The examples that I have given are intended solely to alert you to the numerous footnotes that can sometimes appear on the title page of a manuscript. Additional possible footnotes include references to sources of funding, such as, "Mary Gifted was the recipient of a fellowship from the National Institute of Science, U.S.A.," although sources of funding usually appear in the Acknowledgments at the end of each paper.

3.5 Key words

To facilitate the indexing of published papers, it is generally necessary for the author to supply a list of the key words in his paper. By convention, the term "key words" applies to short phrases, as well as single words, since some terms, such as the names of species, cannot be reduced to single words. Your choice of key words should reflect the most important aspects of your paper. You should choose them in such a way that people interested in your field will find a reference to your paper when they refer to certain words in the journal's annual index. Some journals require key words that do not appear in a paper's title; in other cases, words that are prominent in the title should be used as key words, together with important terms from the manuscript itself if the number of key words that are allowed is sufficient to include these additional terms.

Sometimes, the term "keywords" is used rather than "key words" and, furthermore, the key words (or keywords) may need to be separated by semicolons, long dashes, or short dashes. You need to make sure that you set down your key words in the appropriate format. To draw your attention to the various possibilities, two versions are shown below.

> Key words: *Salmonella typhimurium*—Food poisoning—Hamburger—Refrigeration
>
> Keywords: *Salmonella typhimurium*; food poisoning; hamburger; refrigeration

Note that *Salmonella typhimurium* is not abbreviated to *S. typhimurium* (the names of genera, when they are included in the names of species, are only abbreviated in the main text after they have been written out once in full). The only abbreviations allowed as key words are those that are generally recognized by all scientists in your field and related fields, such as, for example, DNA and RNA in the biological sciences and NMR and EPR in chemistry.

3.6 Abbreviations

The Instructions to Authors usually include a list of abbreviations that can be used without definition as well as instructions about the use of abbreviations that do not occur on that list. Sometimes it is sufficient to define abbreviations as they occur in the text; sometimes a list of abbreviations is required and such a list is often printed after the key words. The Instructions to Authors will tell you exactly where such a list of additional abbreviations should appear in your manuscript, as well as the form that this list should take. Sometimes the list is alphabetical and is printed as continuous text, with appropriate punctuation; sometimes the list is made up of single terms and their abbreviations, one below the other, extending from the top to the bottom of the page. If the list is to be printed as continuous text, you will need to use commas and semicolons properly, as shown in the example that follows.

> Abbreviations: Instructions to Authors, ITA; your target journal, YTJ; and "Letter to the Editor," LTE.

If you are making up your own abbreviations, you need to give them some careful thought. First, you should only make up abbreviations for cumbersome phrases that occur frequently in your text. The best abbreviations are those that are easy to read and assimilate. Moreover, to avoid confusion, your abbreviations should not resemble those that are already widely used in your field. For example, it would be a mistake to abbreviate the fictitious anti-AIDS drug "homoisovenamate" as "HIV" or the fictitious anti-pneumonia agent "specific antagonist of receptors for sputase" as "SARS." You also should avoid abbreviations that are similar to those used in related fields, for example, the abbreviations for the chemical elements. Thus, it would be a bad idea to abbreviate "ligand-insensitive receptors" as "Li receptors" since it could be misinterpreted as "lithium receptors."

It is not a good idea to introduce too many new abbreviations in a single paper because readers will have difficulty following your arguments if they are constantly trying to remember what each abbreviation means. You should adhere to the maxim, "When in doubt, write it out." In other words, if you are not sure whether or not to abbreviate a particular term, write it out in full.

Abbreviations can become useful acronyms that develop a life of their own. A particularly satisfying acronym from the field of immunoassays, for example, is "CANARY," which stands for "cellular analysis and notification of antigen risks and yields" and defines a system for the detection of biohazards, while at the same time reflecting the traditional use of canaries to monitor the safety of the air in coal mines. Unfortunately, authors rarely achieve such a felicitous combination of abbreviation, acronym, and mnemonic and you should not spend too much time trying to do so.

Do not forget, while writing your manuscript, that it is best to avoid beginning a sentence with an abbreviation. Under some circumstances, for example, when you or others have used an abbreviation so often that it has taken on its own life as a word, you are allowed to break this rule and start your sentence with the abbreviation, provided the abbreviation begins with a capital letter. Thus, you could start a sentence with the abbreviation CANARY or AIDS but not with the abbreviation tRNA or cDNA. Moreover, you should never use an abbreviation as the first word of a new paragraph.

3.7 Fonts

Word-processing programs provide access to a large number of different fonts, with names such as Helvetica, Palatino, Monaco, and New York. They also give you access to a variety of font styles, for example, bold and italic. However, the fact that a variety of fonts and font styles is available to you does not mean that you should use a different one for every item on your title page. Choose a simple and unadorned font, such as Helvetica, and use it for every item on your title page. Names of genera (for example, *Homo*, *Pinus*, and *Mus*) and species (for example, *Homo sapiens*, *Pinus radiata*, and *Mus musculus*) should be written in italics, as should words that are not in English, such as "*in vitro*" and "*in vivo*," unless the Instructions to Authors specify otherwise. If the instructions indicate that the title, for example, should be in bold face, you must do as instructed. In the absence of specific instructions, and with the cited exceptions, your entire title page and the rest of the text should be typed in one single font in letters of one single size.

If you are submitting your manuscript electronically or as "camera-ready" hard copy, you must pay very close attention to the font and style that you use. Refer both to the Instructions to Authors and to several examples of papers that have been published in your target journal or on your target website. The use of multiple fonts and styles on the title page is not only very poor style, it is also pointless—the typesetters will usually strip away all your efforts before they start work. It is also unnecessary to "center" the title or the names and addresses of the authors. As Charles Lamb wrote to Samuel Coleridge in 1796, "Cultivate simplicity."

Chapter 4

The Abstract or Summary

4.1 The function and length of the Abstract or Summary

Full-length papers begin with an Abstract or Summary, which I shall refer to exclusively here as an Abstract, for simplicity's sake. Some short communications, such as "Letters to the Editor" in certain journals, do not include an Abstract but, in some journals, even a "Letter to the Editor" requires an Abstract.

The Abstract provides a brief account of the important points in your paper and it allows the reader to judge whether it is worth her time to read the entire text. The Abstract should be written so that it can stand alone, without the full body of the text. Thus, the reader should be able to understand all the material in the Abstract without reference to the main text of the accompanying paper.

The length of the Abstract is generally specified in the Instructions to Authors and you would be wise not to exceed the indicated number of words. If your Abstract is too long, the editor will make you shorten it. Therefore, you should follow the instructions exactly.

In the world of scientists, the term "Abstract" does not apply exclusively to the summary of the research that is described in a scientific paper. It also applies to the summary that a researcher submits when he plans (or hopes) to present his work at a scientific meeting. The Abstract for a scientific meeting can be slightly less formal, in terms of style, than the Abstract of a paper, but the typed Abstract for a meeting generally has to conform to much stricter guidelines with respect to length and layout. These guidelines are provided by the organizers of the meeting and the Abstract has to conform to them in the smallest detail, irrespective of whether the author is a graduate student or an invited speaker. Nonetheless, even if the details of the format, length, and layout of the two kinds of Abstract differ, Abstracts of papers and Abstracts for meetings have one important feature in common: they have to be self-contained and able to stand alone. They should contain no information that can only be understood by reading the accompanying paper or by going to the meeting.

4.2 Heading and numbering

Look at the Instructions to Authors to determine whether the Abstract should typed on a separate page or immediately before the Introduction. You are now, probably, on the second page of your manuscript—the first page was the title page—and now would also be a good time to check how the pages of your manuscript should be numbered and whether the lines of text should also be numbered. You should determine whether the Abstract is labeled and whether it is typed in boldface, as is sometimes the case, or as plain text.

4.3 Format: continuous text or specified sections?

The standard format for the Abstract in most journals is continuous text, with only the length defined. In some journals and, in particular, in medical journals, the Abstract is divided into prescribed sections, such as Background, Methods, Results, and Conclusions. You must use the format imposed by your target journal and make sure that none of the sections is longer than specified.

4.4 Abbreviations

As noted at the beginning of this chapter, the Abstract should be written so that it can stand alone, without any reference to the full body of the text. The reader should be able to understand all the material in an Abstract without reading the main text of the paper. Thus, there should be no unusual abbreviations in the Abstract that are not explained in the Abstract itself. Inclusion of abbreviations that are in common use in your field is acceptable and these abbreviations, which can be used without definition, are often listed in the Instructions to Authors. If the acceptable abbreviations themselves are not listed, the Instructions to Authors usually include references to websites or books where you can find the information that you need. For example, the website of the *American Journal of Physiology* provides a link to a long list of abbreviations in the biological and medical sciences, namely, http://www.the-aps.org/publications/journals/abbrv.pdf. These abbreviations can be used in papers submitted to the *American Journal of Physiology* without definition. For word usage, symbols and other useful information, the instructions refer authors to the book *Scientific Style and Format: The CBE Manual for Authors, Editors, and Publishers* (published by Cambridge University Press in 1994). For chemical and biochemical terms and abbreviations, authors are advised to consult the recommendations of the IUPAC-IUB Combined Commission on Biochemical Nomenclature, which are available on the internet at http://www.jbc.org/cgi/reprint/241/3/527.pdf.

While all this information pertains specifically to the *American Journal of Physiology*, you will find similar information and advice in the Instructions to Authors of your target journal.

With the exception of the abbreviations that you are allowed to use without definition, you may only include in the Abstract those abbreviations that you define within the Abstract. You should also bear this point in mind when you are preparing an Abstract for a scientific meeting. In this case, you should only use abbreviations that you are sure everyone who is likely to attend the meeting will understand without any difficulty.

4.5 The single-sentence summary or précis

When the Abstract is divided into sections, as described above, the author is often required to provide a single-sentence précis of the work in his paper. If the title has been composed as a single declarative sentence, it can often be expanded slightly, by inclusion of the purpose of the study, the methodology or the conclusion, to yield an acceptable précis. Thus, if the title of the paper is, for example, "Refrigeration fails to prevent food poisoning by hamburgers contaminated with *Salmonella typhimurium*," the précis can be written as follows, "As part of an ongoing study to identify methods for the safe storage of ground beef, we found that refrigeration failed to prevent food poisoning by hamburgers contaminated with *Salmonella typhimurium*." Alternatively, if the précis is to include a mention of methodology, it might read as follows, "Using ground meat from a variety of commercial sources and standard conditions for the culture of bacteria, we showed that refrigeration failed to prevent food poisoning by hamburgers contaminated with *Salmonella typhimurium*." If the précis has to include some mention of the conclusion of the study, the title, with slight modification, can be extended as follows, "Refrigeration failed to prevent food poisoning by hamburgers contaminated with *Salmonella typhimurium* and, thus, new methods, such as irradiation, should be examined as part of current efforts to improve the safety of the food supply." These examples also demonstrate the versatility of a carefully composed title.

4.6 Inclusion of references in the Abstract

In general, it is better to avoid the inclusion of references in an Abstract since the Abstract should be able to stand entirely alone, without reference to other materials. Moreover, if you cite previously published papers appropriately in the main body of the paper, it is usually unnecessary to include references in the Abstract. If it is absolutely essential to refer to a previously published paper in the Abstract, you must provide the full reference to that paper in the Abstract even when you cite the reference in the main text of your paper; for example, "Confirming the results of the groundbreaking analysis of the effects of banana skins on the frequency of falls in an urban environment by Berle *et al.* [*Journal of American Pratfalls*, vol. 16, pp. 123–126, 1966], we have shown that . . ." As I have emphasized above, your Abstract should be able to stand alone. Thus, it is insufficient to refer, in your Abstract, to a paper that is included in your list of references or as literature cited at the end of your paper.

For the exact format of references in an Abstract, try to find a complete reference in the Abstract of a published paper in your target journal. If rapid perusal of several issues of the journal fails to locate one, try to write the Abstract in such a way that a reference is no longer necessary, for example, "We have confirmed a previous report that banana skins increase the frequency of falls in an urban environment and extended this earlier finding by showing that . . ." As a general rule, if you are not sure whether what you are writing is correct or acceptable, reconfigure your sentence in such a way that you have no doubts as to its correct structure and appropriate content.

If you are writing an Abstract for a meeting, you can certainly include references, using the same format as indicated above. However, you may find that references take up so much of the limited space available to you that there remains insufficient space in which to summarize your research. Thus, you would be well advised to concentrate on describing your research and to omit any mention of previously published papers until you make your presentation.

4.7 The content of the Abstract

Do not try to include everything that is in the main text of your paper in your Abstract. It is rare that an Abstract is too brief. If your target journal allows 250 words, aim for slightly more than 200 in your first draft. Then you will have a little leeway when you start to fine-tune what you have written. As you write the Abstract, keep the title of your paper in front of you. The Abstract should explain very concisely (i) why you did the study that led to the statement in your title, (ii) how you did the study, (iii) what you found, and (iv) what your results mean. The function of the Abstract is given in Section 4.1 and bears repeating here:

> The Abstract provides a brief account of the important points in your paper and it allows the reader to judge whether it is worth her time to read the entire text. The Abstract should be written so that it can stand alone, without the full body of the text. Thus, the reader should be able to understand all the material in the Abstract without reference to the main text of the accompanying paper.

When your Abstract is complete, compare it with the title of your paper. If the words and phrases in your title are not in your Abstract, you must rewrite one or the other. Your title is the advertisement for your paper. If your title and your Abstract do not correspond to each other, you can consider yourself guilty of false advertising! Indeed, everything you have done up to this point is similar to advertising a product. In a series of advertisements, namely, your running title, your title, and your abstract, you have provided increasing amounts of information to entice the reader to "buy" your product, in other words, to read the paper that you are now ready to write.

Chapter 5

The Introduction

5.1 Length

The Introduction to a full-length paper should be sufficiently long to allow you (i) to place your research in the context of earlier relevant work by others, (ii) to explain your reasons for performing your study, (iii) to mention the methods that you used in your study, and (iv) to provide an indication of the conclusions that you will draw from your results in the Discussion at the end of your paper. However, you need to avoid writing an Introduction that is too long. A long Introduction is appropriate for a dissertation (also known as a thesis) because the doctoral candidate needs to show that she has a very thorough grounding in her field and a full understanding of its history, but the Introduction to a research paper is not intended to show the extent and depth of your knowledge of the field. The researchers who are likely to read your paper are unlikely to be novices in your field and they will not need to be led through its entire history since the earliest experiments by, for example, an obscure Frenchman in the late 1880s. The purpose of the Introduction is to allow those who are at least somewhat familiar with your area of research to orient themselves and to prepare themselves to follow your train of thought, your experiments, and your conclusions from them.

If your Introduction is more than two-thirds the length of your Results section, it is probably far longer than necessary. If the reviewers of your paper complain that your Introduction is too brief, you can always lengthen it. A much more common problem is an Introduction that is too long and needs to be shortened. If you are submitting a "Short Communication" or "Letter to the Editor," your Introduction should be one or two paragraphs at most. Check your target journal to determine whether or not the Introduction requires a subheading.

In many journals, the Introduction begins without its own subheading, unlike the subsequent sections, such as Materials and Methods, Results, and Discussion.

5.2 References in the Introduction

Your description of the recent progress in your field that provided the foundation for your own research will include references to the work of others. Some authors fail to give credit to those on whose work theirs is based. By contrast, some authors are so eager to give credit to everyone who ever made a contribution that the references cited in support of a particular point continue for five or six lines. You should cite the recent papers from the past couple of years that form the immediate basis for your experiments but, rather than lines and lines of references, you should use some version of the all-encompassing phrase, "Brainy *et al.* (2007) and references therein." Using this format, you are able to refer the reader to all the references cited by Brainy *et al.* in 2007 and, thus, to all the relevant work published before Brainy *et al.* wrote their paper in 2007.

The format for citing the works of others is specified in the Instructions to Authors of your target journal. Some journals use numbers in parentheses, such as (1), (1, 2) and (1, 2, 3), for example, ". . . as described by Smart and Gifted (1, 2, 3)," while others use numbers as superscripts, for example, ". . . as described by Smart and Gifted[1,2,3]." Note that, when numbers are given in parentheses, there is a space after each comma, as in the case of commas between words in the text; in the case of superscripts, there are no spaces. In some journals, the superscripts extend beyond the punctuation, for example, ". . . as described by Smart and Gifted,[1,2,3] in their most recent reports;" in some journals they do not: ". . . as described by Smart and Gifted[1,2,3], in their most recent reports." You should check these details in papers that have already been published in your target journal and be sure to use the correct version. If the references in the journal are given as the authors' names in parentheses, you need to check (i) how many names should be included in each reference, (ii) whether or not "*et al.*" is in italics, and (iii) whether the punctuation includes commas and semicolons. I should mention here, for those authors who never studied Latin, that "*et al.*" is an abbreviated version of "*et alii*," which means "and other people." Thus, strictly speaking, "*et al.*" should always be in italics, as should other Latin words, such as "*in vitro*," "*in vivo*," and "*in vacuo*" (in a test tube, for example; in living cells or organisms; and in a vacuum). However, some journals do not italicize "*et al.*" and some do not italicize any Latin words. Therefore, you should always check the Instructions to Authors for instructions about italicization. Whether "*et al.*" is italicized or not, if you remember that "*al.*" is the abbreviation of "*alii*," you will never forget the period (or full stop) that is required.

Some possible formats for the citation of references are given in the following examples:

> . . . as discussed elsewhere (Smart et al. 1999, Brainy et al. 2007).
>
> . . . as discussed elsewhere (Smart et al., 1999; Brainy et al., 2007).
>
> . . . as discussed elsewhere (Smart and Gifted, 2000; Brainy, Gifted and Smart, 2007).
>
> . . . as discussed elsewhere by Smart and Gifted (2000) and Brainy *et al.* (2007).
>
> . . . as discussed elsewhere (Smart *et al.* 1999, Brainy *et al.* 2007).
>
> . . . as discussed elsewhere (Smart *et al.*, 1999; Brainy *et al.*, 2007).

The differences among the examples shown above are very small but you need to be sure to use the exact format that is required by your target journal.

If you are using a numerical format in the citation of references, be sure that the first reference that you mention in the text is number one (1). Readers find it quite perplexing when the first reference mentioned in the Introduction is, for example, number eight (8) but you would be surprised how frequently I correct such an obvious mistake.

5.3 Historical background

As noted above, you do not need to provide a complete history of your field in your Introduction nor do you need to cite every paper that was ever written in your field. Authors who have not yet made a name for themselves look eagerly for references to their work and for their names in the works of others, while authors whose recent work has made a seminal contribution to the field will find their names mentioned frequently. By contrast, authors whose work has made them very famous accept that their names no longer appear and that their groundbreaking discoveries have become accepted as facts. Thus, for example, most papers on DNA no longer include a reference to the paper in which Watson and Crick described their discovery of the double helix. The double-helical structure of DNA is an accepted fact and no reference need be made to its discoverers in a standard research paper. By omitting references to discoveries that are accepted as facts and using the phrase ". . . as described by Brainy *et al.* (1999) and references therein," you should be able to document relevant advances in your field without taking up too much space. You should be aware, moreover, that the reviewers of your manuscript will be researchers in your field so you must be careful to document recent progress in your field judiciously.

5.4 The working hypothesis behind your research

The results that you are going to describe in your paper are either the results of experiments that you performed to test a working hypothesis or they are the serendipitous results of experiments that you planned with an entirely different goal in mind. In both cases, you need to lay the groundwork for your experiments in your Introduction. However, if your results were serendipitous, you should not dwell at length on what you had hoped to discover when you began your study but you should present the background that places your unexpected results in an appropriate context.

5.5 Methodology, instrumentation, materials and analytical tools

If, in your study, you applied or developed novel methods or instrumentation or used uncommon or unusual reagents or materials, you should mention in the Introduction how and why you did so. You should also mention the application of novel or non-standard analytical tools, again providing your reasons.

5.6 Relevance of your study and inferences from your results

You should end your Introduction by mentioning the way in which you will discuss your results in the Discussion and place them in a wider context. You should not, however, fall into a common trap by writing, for example, "Based on our results, we shall present a model of the detailed dynamics of the complex photochemical reaction." This incorrect construction indicates that "we" are "based on" the results, whereas it is actually the model that is based on the results. You should say, "In the Discussion, we shall present a model of the detailed dynamics of the complex photochemical reaction that is based on our results."

The advice in this chapter can also be applied to the introductory paragraphs of a "Letter to the Editor" and a "Short Communication" but, in these cases, the introductory paragraphs should be much briefer than the Introduction to a full-length research paper.

Chapter 6

Materials and Methods

6.1 Format

The section in which you describe how you performed your experiments generally comes after the Introduction. In some journals, however, the methods are described at the end of the paper and, in some journals, the section that includes details of materials and methods is called "Experimental Procedures."

The purpose of the Materials and Methods or Experimental Procedures section of your paper is to provide information in sufficient detail that another scientist in your field should be able to repeat your experiments and reproduce your results. This section also allows the reader to judge whether you used the appropriate materials and instrumentation, as well as the best techniques, to obtain your results. Before you begin, check the format in the Instructions to Authors. Pay particular attention, when you consider the format, to the answers to the following questions. Should the title have one capital M or two ("Materials and Methods" or "Materials and methods")? Is the title in the center of the page or on the left. Are the subtitles in plain text, boldface or italics? (If the subtitles are in italics, any words in the subtitles that would normally be in italics in plain text are written in plain text instead of italics, for example, "*Determination of light sensitivity* in vitro.") Are the nouns in each subtitle capitalized or is only the first word capitalized? Is each subtitle followed by a period (or, as the British say, a full stop), a colon (:), a long dash (-- or —) or a short dash (- or –)? Is each subtitle on a line of its own or does the text follow immediately after and on the same line as the subtitle?

The following examples are all different but your target journal will accept only one format. The journal's format might not be included below but the examples are illustrative in so far as they show how many possibilities exist and how careful you have to be to choose the right one.

> Cells and culture conditions.
>
> The cells used in the present study were . . .
>
> Cells and Culture Conditions. The cells used in the present study were . . .
>
> **Cells and culture conditions**. The cells used in the present study were . . .
>
> **Cells and culture conditions**
> The cells used in the present study were . . .
>
> Cells and culture conditions: The cells used in the present study were . . .
>
> Cells and culture conditions—The cells used in the present study were . . .
>
> Cells and culture conditions – The cells used in the present study were . . .
>
> *Cells and culture conditions.* The cells used in the present study were . . .
>
> *Cells and culture conditions*
> The cells used in the present study were . . .

Consistency of presentation is an important feature of a well-written manuscript and you must use the appropriate format consistently in the Materials and Methods (or Experimental Procedures), Results and Discussion sections of your manuscript.

6.2 Biological samples

If your research involved any biological samples, you should provide details at the start of this section. If you used samples from human subjects or human subjects themselves, you must indicate that you obtained permission from the Human Investigations Committee (or the equivalent) of your institution. If your institution does not have such a committee, you should follow and refer to the guidelines in the World Medical Association's "Helsinki Declaration," which details the ethical principles for all medical research that involves human subjects. You can find this material on the internet at http://www.wma.net/e/policy/b3.htm.

If you used animals, you must give their sources and certify that they were handled and treated according to humane practices, as defined by your institution or source of funding. You should also provide details of the care and treatment of all animals prior to your experiments. If you used cells, you should specify the sources in sufficient detail that another researcher could obtain samples of the same cells if needed. The same holds for plant materials.

After you have described the nature, source and maintenance of your biological samples, you should discuss their treatment.

6.3 Chemicals

It is not necessary to specify the sources of basic laboratory chemicals, but it is necessary to specify the sources of all non-standard chemicals and macromolecules. You should give the full name of each item and, if the name is complicated, now is a good time to introduce an appropriate abbreviation. For each item, you need to specify the manufacturer and the location of the manufacturer. If the manufacturer's name occurs more than once, you should not repeat the location. Here is an example that incorporates all the above advice:

> Sedoheptulose anhydride (SEDAN) was purchased from Sigma Chemical Co. (St. Louis, MO). We also used the following antibodies: monoclonal antibodies against sputase (MabS; Mabco Co., Ltd., Podunk, CA) and against spitase (MabSp; Mabco Co., Ltd.); and polyclonal antibodies against lungulin (prepared as described by Gifted *et al.*, 1998).

Notice, in particular, the use of the semicolon after the abbreviations "MabS" and "MabSp" and the commas that follow within the respective parentheses. If you find yourself unsure of the exact spelling of an item or the location of the manufacturer, you can use a search engine, such as Google (http://www.google.com) to find the information rapidly. If you misspell the name of a compound or supplier, the Google search engine will probably recognize it anyway and suggest the correct spelling, allowing you to locate information about the compound or about the supplier.

6.4 Units of measurement

As you start to describe your methods, you need to check that you are using the right format for units of measurement. Most journals require that you use the SI metric system. The abbreviation SI stands for *Système International d'Unités*, which means the International System of Units. In this system, with which you are doubtlessly familiar, the base quantities for length, mass, time, electric current, amount of a substance, and temperature are the meter (m), kilogram (kg), second (s), ampere (A), mole (mol), and degree Kelvin (K). You can find additional details of this system at http://physics.nist.gov/cuu/Units/. In addition to using the correct units, you need to determine whether to use a slash (/; also known as "solidus") or a superscript in place of the spoken "per." Thus, while you would say, "grams per liter," you must decide among "g/l," "g l^{-1}" and "g.l^{-1}." In general, the last of these three versions is preferred because it allows complex units to be described simply and clearly. For example, "micromoles per meter squared per second" is more clearly abbreviated as "µmol.m^{-2}.s^{-1}" or as "µmol m^{-2} s^{-1}" than as "µmol/m^2/s."

To avoid common problems associated with the capitalization of geochronologic and chronostratigraphic units, geologists should refer to an authoritative explanation of the appropriate and correct usage of stratigraphic terminology at http://www.agiweb.org/nacsn/JSP_commentary.htm.

No matter which format or units you choose or are instructed to use, it is very important to be consistent and to use the same format and respective units throughout your manuscript, in your Figures, and in your Tables.

6.5 Registered trademarks

If you use any materials or instruments whose names are registered trademarks, you should include the superscript TM or ®, as indicated in the printed material from the manufacturer or supplier.

6.6 Organization of the Materials and Methods

Once you have dealt with the sources of the materials that you used for your research, you must describe the methods that you used. These methods fall into four groups: methods with which everyone in your field and related fields is familiar; methods that are in less common use but have been well documented elsewhere; methods that are relatively uncommon or that require specification of experimental conditions for each application; and novel methods that you developed for the research described in your paper.

Methods in the first group, for example, the method for determining the pH of a solution, need not be mentioned. Methods in the second group, such as the quantitation of chlorophyll in an extract of leaves or of protein in an extract of cells, should be mentioned briefly with the appropriate reference and an indication, when necessary, of the materials used for standardization of the method, for example, "Proteins were quantitated as described by Bradford (1976) with bovine serum albumin as the standard protein." Methods in the third group should be described in sufficient detail that someone who wants to repeat your experiment can do so by referring to the original description of the method and to the specific conditions that you used. Thus, such methods are written up according to the following formula, "The gene for sputase was amplified by the polymerase chain reaction (PCR), as described by Brainy *et al.* (2007) for the amplification of the gene for spitase, with the following modifications." Methods in the fourth group should be described in detail, with all reagents, conditions, and equipment carefully specified. As you write the description of your novel method, you should also indicate when and how special care should be taken to ensure success. For example, "It is essential to add solution B immediately after solution A has been added to the mixture. Any delay will result in the formation of a precipitate and activity will be lost."

The order in which you describe your methods should, to some extent, reflect the order in which you used your methods to obtain your results. In general, after you have indicated the sources of your samples and reagents, you will discuss experiments with these samples and reagents, the methods used for the analysis of the products of these experiments and, finally, the methods used for the analysis of your results. It is important to be very specific about the methods that you used to analyze your results. In particular, you need to specify the size of any sample that you analyzed and the number of times that you repeated each of your individual experiments. You should indicate whether the results that you show are means with standard deviations (or standard errors of the mean) of the results from a particular number of samples or examples of results obtained from a specified number of replicate experiments, all of which gave similar results. If you subjected your data to statistical analysis, you need to include the statistical methods that you used, as well as the criteria that you used to establish the significance of differences between results (for example, the "*p* value," if you performed your statistical analysis using Student's *t*-test).

The Materials and Methods or Experimental Procedures section is generally organized along the lines described above. You should, however, check the Instructions to Authors, as well as some published papers in your target journal, to determine whether you need to modify the approach that I have suggested to satisfy the requirements of the journal.

We should not forget, here, an error that occurs very frequently in efforts to summarize methods that have been described by others. An author, wishing to be concise, begins his compressed description of a method with the word "Briefly." For example, "Briefly, the data were subjected to statistical analysis, as described by Gifted *et al.* (2007), and statistically significant correlation coefficients were recorded." The adverb "briefly" modifies the verb "were subjected" and does not reflect the author's intention to be brief. Instead of "Briefly," the first words in the sentence should be "In brief," as follows: "In brief, the data were subjected to statistical analysis . . ."

6.7 Details of theoretical premises and computations

If your study involved complicated theoretical premises and computations, you should include, at this point, the theoretical background and computational details of your study plus an overview of your calculations, if such an overview is appropriate.

Chapter 7

Human subjects

7.1 Descriptions of human subjects and case histories

If your study involved human subjects, you should discuss their relevant features in the Materials and Methods section under the subheading "Human subjects" or "Patients," or in a separate section, similarly entitled "Human subjects" or "Patients." Moreover, if the only biological "materials" in your study were human subjects or patients, you can use the heading "Human Subjects and Methods" or "Patients and Methods" instead of "Materials and Methods." This format is appropriate if the subjects or patients who were included in or who participated in your study formed one or several basically homogeneous groups, such as males between 35 and 45 years of age or patients with specific malignancies at various defined clinical stages. By contrast, if you studied a small number of patients whose clinical picture was non-standard or whose respective conditions led you to pursue a particular line of questioning or hypothesis, you should provide a separate case history for each patient. You should never include any information that might allow someone to identify your human subjects, for example, their names, initials, or hospital identification numbers. Pay attention also to identification numbers on radiologic films and be sure to mask faces in any photographs of human subjects.

7.2 Informed consent

Whenever you study people or tissues from people, you need to obtain the formal and informed written consent of your subjects (or their parents, if they are minors) or you need to determine unequivocally that such consent is not required. At many institutions there is a Human Investigations Committee and, if you are in any doubt whatsoever about the need to obtain informed consent from the subjects of your proposed study or experiments, you should submit your research proposal to this committee before you start your research. If informed consent was required before you began your study, you should indicate, at first mention of the nature of your study, that you obtained such consent. For example:

> In accordance with the regulations set forth by the Human Investigations Committee of the University of Erewhon, informed consent was obtained from all adults and from a parent or guardian of all subjects of less than 18 years of age.

As noted in Chapter 6, if your institution does not have a Human Investigations Committee, you should follow, during your study, the guidelines in the World Medical Association's "Helsinki Declaration," which details the ethical principles for medical research that involves human subjects. You can find this material on the internet at http://www.wma.net/e/policy/b3.htm. If you have followed these guidelines rather than those of your institution, you should mention the "Helsinki Declaration" and confirm your strict adherence, during your study, to the ethical principles laid out in this document.

7.3 The format of a case history

In general, each individual case history is presented separately and each is printed as continuous text without subheadings. As noted above, it is essential that you protect each person's privacy; you should never include a person's name, initials, or hospital identification number for any reason whatsoever. You can assign each patient an identification number for the purposes of your study but such numbers should not allow anyone to identify individual patients.

Each case history should flow from the simple to the complex, reflecting, as far as possible, a chronological sequence of events. The first sentence of a case history always includes a description of the patient in terms of age, gender, and race or nationality plus the issue that first brought the patient to the attention of physicians, for example, "A 35-year-old Japanese woman presented with a cough of two years' duration." The next sentences should describe any relevant aspects of the patient's own medical history or of her relatives' medical history, including any genetic abnormalities and details of consanguineous marriages. The history should continue with details of the patient's medical condition and the various tests, such as radiologic examinations, laboratory tests, and molecular biological tests, that were performed in an attempt to make a diagnosis. The results of these tests should also be included here if there is no separate Results section in your paper.

The case history should continue with a description of the patient's medical treatment, together with the details of any relevant tests that were performed during her treatment either to monitor the effects of the treatment or to provide additional diagnostic information. Again, if there is no separate Results section, the results of these tests should be included here.

The case history should end with details of the outcome of the treatment, the patient's status at the time of writing, and any plans for specific follow-up by medical professionals.

Some short papers in medical journals consist of one or more case histories that are followed immediately by a Discussion. In papers that include the experimental treatment of patients or the analysis of specimens obtained from patients, for example, blood samples or tissues removed at biopsy or autopsy, the case histories are included after the Materials and Methods or Experimental Procedures section, which should include details of all analytical methods, and before the Results section. Thus, if you are describing a genetic analysis of six patients that involved amplification of genes by the polymerase chain reaction (PCR), you should describe the details of PCR in Materials and Methods or Experimental Procedures, and then you should provide the case histories of each of the six patients in a subsequent section.

The formats of all medical journals are not identical. Therefore, to ensure that you are using an appropriate format, you should model your presentation on similar papers in your target journal.

Chapter 8

Results

Before you start to write the Results section, pause for a moment to remember its purpose. In this section, you will lead the reader along a path that follows the steps in your reasoning and the sequence of your experiments from your working hypothesis (or definition of the purpose of your study) to a conclusion that corresponds exactly to the title of your paper.

8.1 The quality of your data

You should only publish experimental results that are absolutely reproducible. You should ensure that your results are reproducible by performing each experiment several times, with several replicates within each experiment, if possible. Your paper serves not only to communicate your findings to others but also to allow others to repeat your experiments and build on them. Thus, if you yourself cannot obtain reproducible results, it is unlikely that anybody else will be able to do so and, for that reason, your results are not worth publishing. The validity of your results also depends on the size of your sample: the larger the sample, the greater the value of your results. If your results are not reproducible, do not despair. There is a place for them too, namely, the *Journal of Irreproducible Results*, which is happy to publish "news of particularly egregious scientific results." If your findings fall into this category, you can submit them to the journal via the its website at http://www.jir.com/home.html.

8.2 What results should you include in your Results section?

You should only provide results that pertain directly to the title of your paper. If you have negative results that are relevant, it is often appropriate to mention them with the phrase, "results not shown" or "data not included" in parentheses. References to unpublished results, which should be kept to a minimum, should be followed by the phrase, "our unpublished results," or, for example, "unpublished results of M. Tinker and B. Bell" or "personal communication from M. Tinker" and each of these phrases should be in parentheses. If your study involved the accumulation of vast amounts of data that cannot be included in a regular paper, you can provide a reference to a website where they will be available. This website might be a site on which your target journal prints supplementary materials; it might be a website where all results such as yours are collated; or it might be a site that you set up for the sole purpose of displaying your data. You should check, however, to determine whether your target journal allows references to personal websites.

8.3 The organization of your results

Unless your paper is very brief, you should organize your results under subheadings and the format of the subheadings should be exactly as specified by the Instructions to Authors of your target journal. The format should also be exactly the same as the format that you used in the Materials and Methods or Experimental Procedures section. If your paper is very brief, for example, a "Letter to the Editor" or a "Short Communication," your entire text might not include any subheadings. Look at your target journal for guidance in such cases.

Some journals allow or encourage the condensation of Results and Discussion into one section entitled "Results and Discussion." In this case, you can discuss the importance of each result after you have presented the relevant data, finishing your entire paper with a few concluding sentences in which you summarize your study, its importance, and the possible or actual direction of future research.

8.4 Presentation of your data

You should present your data in a logical sequence, with each individual experiment or series of closely related experiments under its own subheading. Whenever possible, you should provide illustrations, that is to say Figures, that support your descriptions of your results, for example, photographs, micrographs, histograms (bar graphs), and graphs. You should also provide Tables when you have more data than you can summarize easily in the text. You should avoid, however, including Tables that contain no information other than information that is already in the text. For example, if you include in your text the observation that the electrophoretic mobilities of three proteins, A, B, and C, indicated that their molecular masses were 24 kDa, 75 kDa, and 32 kDa, respectively, you do not need and should not include a Table, with the title, "Molecular masses of proteins A, B, and C," that includes only those values. If you are going to present your data at a seminar, you might make use of such a Table but it would be a waste of space in a journal.

8.5 References to Figures and Tables in the Results section

Confucius may or may not have said that one picture is worth a thousand words but there is no doubt that illustrations can convey information more succinctly than lengthy passages of prose. When provided as part of your results, photographs and photomicrographs provide instantly assimilable information, while graphs and histograms allow the easy interpretation of results that involve the dependence of a particular parameter on a given variable or set of variables. You should be able to tell a great deal about the appropriate number and style of your Figures and Tables from looking at published papers in your target journal and at the Instructions to Authors. If you have large amounts of data that you feel that you must publish to support your conclusions, you might be tempted to cram numerous minute reproductions of graphs, histograms, and photographs into a single Figure, with designations such as "Figures 1Aa, 1Ab, 1Ba, 1Bb, 1Ca, 1Cb, and 1Cc." Journals allow a certain amount of leeway in this regard but, if the reader can only make out the text, symbols, and relevant details in your Figures with a magnifying glass, you should simplify your Figures. You will find additional information about the preparation of Figures, Legends to Figures, and Tables in Chapters 12 and 13.

Within the Results section, you should refer to each Figure immediately after you describe the result that is illustrated in that Figure. Moreover, for submissions to certain journals, the instructions specify that you should indicate where, in the text, each Figure and Table should be printed. Such annotations are usually written in pencil, in the margin of your text, when you are submitting "hard copy," that is to say, a manuscript printed on paper.

Many scientists find it helpful to prepare their Figures and Tables, using the material from their notebooks or computer files, before they start writing the text of their papers. Then they organize their written material around these Figures and Tables. This strategy is certainly appropriate if each of your results is illustrated by a Figure or a Table. If your paper includes the results of complex calculations or the complicated analysis of data, you can adopt a similar strategy, building your text around your calculations or analysis.

8.6 The commonest mistakes in the Results section

The commonest mistakes in the presentation of numerical data involve the number of significant figures and decimal places. For example, if you have measured a length with a meter ruler, your results may be accurate to the millimeter but not to the micrometer. The problems with decimal places and numbers of significant figures generally arise because inexperienced scientists perform calculations with a calculator or a computer. For example, when I use my computer's simple calculator to determine the average molecular mass of three proteins, whose electrophoretic mobilities indicate that their individual molecular masses are 118 kDa, 71 kDa, and 62 kDa, my calculator produces the result 83.666666667 kDa. Since the measurements that gave rise to this average were not accurate to nine decimal places, the average cannot be accurate to nine decimal places. The meaningful average value of the molecular masses is 84 kDa.

If you are in any doubt about the number of significant figures or decimal places that you should use, think carefully about your results, the message that you are trying to convey and whether the number of significant figures or decimal places that you want to include in a specific numerical result conveys an appropriate message. Remember also that, when you are extrapolating from a graph, the numbers that you extract depend on the way you (or the computer) drew the line through your data points. If the line does not go directly through every point, you or the computer had to choose an appropriate line. This line is, thus, to some extent theoretical and the numbers that you derive by extrapolation from the line are theoretical. They are certainly not accurate to more significant figures than your original data. There are rules of thumb, such as "when you add or subtract, you should retain as many decimal places as there are in your least accurate measurement; when you multiply or divide, you should retain as many significant figures as there are in your least accurate measurement," but I prefer the rule, "always look at your numerical results intelligently and limit the number of digits in each to the number that is meaningful."

8.7 Availability of your newly synthesized materials to others

Your decision to publish your paper in a particular journal means that anyone is allowed to repeat your experiments and to take advantage of them in any way that he chooses. By publishing, you are also agreeing, in many cases, to provide any specialized reagents that you generated during your research, for example, antibodies, lines of cells and plasmids, and chemical compounds, to other scientists, provided that the latter do not plan to use these reagents for commercial purposes for which you have already obtained a patent. This stipulation is included in the Instructions to Authors of some journals and you should pay careful attention to it if you might be unwilling to share your newly synthesized materials. When you describe the preparation and properties of novel reagents in your results, you may become obligated to share them with other scientists within the limits of practicability and feasibility.

8.8 Intellectual property and patents

If your paper contains a description of some invention, material, or process for which you hope to receive a patent, you need to be aware that, in order to obtain valid patent protection in the United States, your patent application must be filed with the Patent and Trademark Office of the United States within one year of the date on which a description of your invention, material, or process appears in print for the first time, anywhere in the world, and within one year after the invention or product first goes on sale or into public use in the United States.

In this context, you should also be aware that not only publication in print but also an oral presentation at a scientific meeting or to a company constitutes public disclosure. The Patent Office of the United States gives you one year from the date of written or oral disclosure during which you retain the rights to a domestic patent. However, you lose all rights to any foreign patents automatically when you disclose your invention to the public.

Submission of a manuscript to a journal, as distinct from actual publication, is generally not considered the equivalent of disclosure. However, if you are following your dates carefully, you need to pay attention to the fact that your paper might appear on a journal's website before it appears in print. If you ignore the deadlines noted above, you might find that your invention is no longer considered "new" for patent purposes and is, therefore, not patentable.

Laws, rules, and regulations are subject to change both in the United States and elsewhere. Thus, you would be wise to check with the appropriate office at your institution or company to determine whether you need to protect your invention before you publish your work or discuss it publicly.

You should include no discussion of your results and no conclusions from them in the Results section beyond a brief mention of how one result led you to perform the experiment that led to the next result. You do not need a summary sentence or a list of conclusions at the end of the Results. However, if you present your results in a section entitled "Results and Discussion," you should discuss each result individually after you have described it and then, at the end of the entire section, you should briefly discuss the conclusions that can be drawn from all your results together.

Chapter 9

Discussion

9.1 Length and purpose

Reviewers rarely complain that the Discussion section of a paper is too brief. If your Discussion is longer than your Introduction or your Results, it is probably too long. You need to resist the temptation to overanalyze and overinterpret your results. If your Discussion is too short, you can be sure that you will receive a request to lengthen it from the editor or reviewers of your target journal. In such cases, the reviews of your paper will probably contain very specific instructions about any additional information or analysis that you should include in the revised version of your Discussion. It should be easy to satisfy such requests when you resubmit your paper. Your Introduction should have provided a clear context for your experiments and related them appropriately to recent progress in your field. There is no need to repeat what you have written in the Introduction beyond a single introductory sentence that places your results in an appropriate context.

The purpose of your Discussion is to provide a summary of each of your results and to show the reader how these results led you to the conclusion that corresponds to the title of your paper. The reader should be able to follow each step in your reasoning and, referring to your results as you summarize them, she should find your Discussion so persuasive that she inevitably reaches the same conclusions as you do. As I noted earlier, scientific truths are rare and, even if you are able to convince the reader that your conclusions are valid, you should be aware that your interpretation of your results might not stand the test of time. It is for this reason that your results must be absolutely reproducible.

Your main responsibility in publishing your work is to add reproducible results to the body of scientific knowledge. Your Discussion represents your best effort to interpret your results appropriately in the context of the current state of scientific knowledge and understanding. It may be reasonable to interpret your results in several ways but you should not try to use your results to lay claim to vast areas of future research by predicting numerous possible outcomes of future work in your field. It is always useful to remember, in this context, the terse statement at the end of the "Letter to *Nature*" in which Watson and Crick first proposed the double-helical structure of DNA, namely, "It has not escaped our notice that the specific pairing we have postulated immediately suggests a possible copying mechanism for the genetic material" (*Nature,* 1953, volume 171, pp. 737–738). This single sentence, written more than 50 years ago, encapsulated the essence of all subsequent research in molecular biology. It is also worth remembering that, in their "Letter to *Nature*," Watson and Crick used fewer than one thousand words to announce and discuss their groundbreaking discovery.

9.2 Organization of the Discussion

The Discussion should begin with one or, at most, two introductory sentences, in which you allude briefly to the current state of knowledge in your field and to your working hypothesis, if you had one at the start of your study, or to the goal of your study. You should discuss each of your results in the same order as you presented them in the Results section. You should decide whether the discussion of your results will be clearer with or without subheadings. If you use subheadings, you should use the same format, in terms of, for example, font, capitalization, and italicization, as you used in your Materials and Methods and Results sections.

You should not introduce new results from your study in the Discussion. Moreover, if you mention the preliminary results of experiments that you have begun since you finished the work that you are discussing, the reviewers of your paper are quite likely to request that you complete your preliminary experiments and include the results in a revised and expanded version of your paper.

The discussion of each of your results should lead readers through a logical sequence that begins with your working hypothesis, if you had one, or your goal and ends with the conclusion that is spelled out in the title of your paper. In the final paragraph of your Discussion, you should consider the present and future impact of your results and of the conclusions drawn from them. You might choose to discuss the practical implications, if any, of your research and/or you might mention the experiments that you plan to perform or are currently performing to extend your results and amplify their importance.

As you write your Discussion, try to keep it as concise as possible. If you follow the guidelines and the advice that I have given you, there should be little extraneous material in your Discussion and it should not be too long. If it is longer than the Materials and Methods section or the Results, you should be able to abridge it by shortening the discussion of each individual result or, perhaps, by discussing only your major results, while alluding briefly to less important results. As I noted at the beginning of this section, if the reviewers of your Discussion consider it to be too brief, you will have the opportunity to lengthen it when you revise your text according to their suggestions.

Chapter 10

Acknowledgments

10.1 The purpose (and spelling) of Acknowledgments

The Acknowledgments section of your paper is the place where you can and should thank all those people who are not listed as authors but whose efforts contributed to the research described in your paper and to the preparation of the paper itself. You should also acknowledge all the organizations that contributed financially to your study. Occasionally, each source of funding is included in a footnote to the title of a paper and you should check your target journal to determine where to put your acknowledgments of sources of funding.

The spelling of both "Acknowledgments" and "Acknowledgements" is correct; you should use the version preferred by your target journal.

10.2 Who gets acknowledged?

The Acknowledgments section should begin with the words, "The authors thank . . ." Any more complicated formula, such as, "The authors would like to express their gratitude to . . ." is unnecessary. Moreover, it is better to begin this section with "The authors thank . . ." than with "We thank . . ." or with "The author thanks . . ." rather than "I thank . . .," if there is only a single author. By contrast, in a master's thesis or doctoral dissertation, the author should use the first person singular when expressing his gratitude to his supervisor, his colleagues and others and even, as I have seen on occasion, his cat.

It is customary to thank people in the following order: people who provided intellectual input that formed part of the basis for your study or who encouraged the authors in their work; people who assisted with some specialized measurements or allowed the use of specialized equipment; people who provided specific materials; people who provided technical assistance; people with whom the authors discussed their results during or after the study; people who commented on the original or revised manuscript; the person who prepared the manuscript (your secretary, for example); sources of funding, with the initials of individual authors included, in parentheses, when funds were allotted to a specific author; and fellowships awarded to specific authors, with the initials of the respective authors, again in parentheses. You should mention the affiliations of all the individuals mentioned, with the exception of technicians and your secretary. Here is an example:

> The authors thank Professor Maria Boffin (University of Podunk Medical Center) for suggesting that they examine the sequence of the gene for lungulin; Professor James Savant (Dept. of Biology, University of Podunk) for his encouragement during the early days of the study; Dr. Elizabeth Mechanic (Instrument Center, University of Podunk) for her assistance with the confocal laser scanning microscope; Mr. Matthew Purchase (Useful Biochemicals, Co., Ltd., Erewhon, CA) for the gift of antibodies against lungulin; Ms. Jenny Yamamoto for her skilled technical assistance; Dr. Daniel Theman and Ms. Leah Hazard for helpful discussions; and Mr. Eli Dean for his patience in preparing the original manuscript. The work described in this report was funded by a grant (to A.B.) from the National Institute of Pneumoscience (USA) and by a grant for "Cooperative Research into Respiratory Diseases" from the Ministry of Science of Japan (no. 12345; to N.T.). T.P. is the recipient of a fellowship from the Eurocentric Association of Science and Technology.

Note, in the example above, the semicolons that separate the phrases about each individual or pair of individuals and the fact that the initials of those who received funding are not separated by a space ("A.B." and "N.T." rather than A. B. and N. T.). You should also remember that you can only use abbreviations that have already been defined in your manuscript. Thus, you must write "National Institute of Pneumoscience" in full, rather than using the abbreviation "NIP," unless you have defined "NIP" earlier in your manuscript.

Abbreviations for names of states and other abbreviations used in everyday life can, of course, be used without definitions.

10.3 Conflicts of interest

A conflict of interest arises when there is a financial relationship between any of the authors of a paper and the material that they have studied. This relationship might include direct or indirect funding for the project from the manufacturer or supplier of the material. Alternatively, you or another author of your paper might have a financial or commercial interest in the manufacturer or distributor of the product. You are obligated to disclose any possible conflict of interest in the letter to the editor that accompanies your paper when you submit it to your target journal. You should also note any possible conflict of interest at the end of your Acknowledgments.

Chapter 11

References and Notes

11.1 References to papers

You must pay very careful attention to the format of your list of references. You will save yourself time and frustration by putting together this section of your paper with infinite care. There are few aspects of the preparation of a manuscript that are more tedious and irritating than correcting a list of references that has been compiled according to an incorrect format. To avoid making mistakes that will require painstaking corrections, study a list of references that has already been published in your target journal in addition to the Instructions to Authors.

The first thing that you should note is the title of this section of your paper. Is it called simply "References" or is it "Literature Cited" (or "Literature cited")? Next, you need to check whether the references are in numerical order, with the order corresponding to the order in which they appear in the text, or in alphabetical order, or even, as is the case in some journals, in numerical and alphabetical order (in this case, the references are listed in alphabetical order and then numbered).

A glance at the lists of references in several journals in different disciplines from different publishing houses is enough to demonstrate that the specified permutations and combinations of punctuation and spacing are numerous. The possibilities range from the simple to the complex, for example, two authors can be listed simply as "Schwartz JA and Weiss S" or as "Schwartz, J.A., and Weiss, S." In some journals, there is no comma after the first set of initials in the second of these examples (even though there should be a comma): "Schwartz, J.A. and Weiss, S." and in some journals, an ampersand "&" replaces "and" to yield "Schwartz, J.A., & Weiss, S." or "Schwartz, J. A. & Weiss, S."

When there are more than two authors, you should use the extended form of one of the above formats (or the specific format of your target journal) but here, too, you must pay attention. Some journals require the name of every author of a paper, even if there are a dozen or more; other journals require that you list a certain number of authors, for example, five, and then follow this list with "*et al.*," which means "and other people." If you are using this format, you should check whether "*et al.*" should be in italics or not. Whatever format you use, you must be sure to use it with absolute consistency. You should also note whether the year of publication comes after the names of the authors or after the title of the paper.

Many journals include the title of every paper in each list of references. When you list the titles of papers, you should double-check them to make sure that you are quoting each accurately. The paper whose title you misquote may be a paper written by the person who will review your paper. She will not take kindly to your carelessness. With the exception of the first word, the words in the titles of papers are generally not capitalized unless they are proper names. Latin words, such as names of genera or species, "*in vitro*" and "*in vivo*," should be italicized unless the Instructions to Authors indicate specifically that italics should not be used. If you refer to a paper in a foreign language, check the spelling very carefully, paying special attention to any accents.

When you have written out the names of the authors of a paper, the title and the date of the paper that you are citing, you need to determine whether to include the entire name of the journal in which the paper appeared or only the abbreviated name. If you are using abbreviated names and are uncertain about the abbreviation of the name of a particular journal, you can find a complete index of such abbreviations at a website provided by the Library System of the California Institute of Technology: http://library.caltech.edu/reference/abbreviations/. Irrespective of whether you are using complete names of journals or abbreviations, you should check whether the name or abbreviation should be in regular type, boldface type, or italics. The same applies to the volume number and the pagination. Pagination usually, but not always, includes the first and last pages of a paper. When you type the reference to the journal, volume number, and pagination, you should continue to pay careful attention to punctuation. Just as there are various ways of listing the names of authors, there are many possible ways of listing the names of journals, the volume number, and the pagination. A few examples are given here:

Spectroscopy in Medicine and Biology, 39, 2130 - 2133

Spectroscopy in Medicine and Biology, **39**, 2130 - 2133.

Spec Med Biol 39, 2130 - 2133

Spec. Med. Biol. 39, 2130 - 2133

Spec. Med. Biol., **39**, 2130 - 2133

Spec. Med. Biol., 39, 2130 - 2133.

Spec. Med. Biol. 39: 2130 - 2133

Spectroscopy in Medicine and Biology 39: 2130 - 2133

Spectroscopy in Medicine and Biology, 39: 2130–2133

Spectroscopy in Medicine and Biology 39: 2130

Spec Med Biol, 39, 2130

Spec Med Biol, 39, 2130.

If the year of publication is not given immediately after the list of authors of the paper, the number of possibilities increases still further, with the year appearing in parentheses or between commas, for example, as follows:

Spectroscopy in Medicine and Biology, 2007, **39**: 2130–2133

Spectroscopy in Medicine and Biology (2007) **39**: 2130 - 2133

Sometimes, a very condensed format is used:

Spec Med Biol (2007) 39:2130-2113

Each of these examples is slightly different from the others and you should look at each to see how it differs from the others. By practicing in this way, you will be better equipped to notice all the details of the specific format that is required by your target journal. You might think that this little exercise is trivial and a waste of time. Let me assure you that, unless you are attuned to the minutia of the requirements of your target journal, you will end up spending much more time correcting a list of references that has been compiled without sufficient attention to detail.

11.2 References to books and to chapters in books

The correct format for references to books and chapters in books requires even greater attention to detail than the format for references to published papers. Once again there are numerous possible versions and you must, again, refer to papers in your target journal and to the Instructions to Authors. In general, all the words in the title of a cited chapter begin with small letters except the first word and words that are capitalized in regular text, such as proper names and the names of genera. By contrast, the titles of books are generally capitalized according to the accepted style for titles, with nouns, adjectives, and verbs being capitalized as they are on the jacket of the book itself. The citation of a chapter of a book should include the names of the authors, the title of the chapter, the title of the book, and the volume number (if any), the editor(s) of the book, the pages on which the chapter begins and ends, the date of publication of the book, and the name and location of the publisher. The order in which these items are listed depends on the journal. Here are a few of the possible formats:

> Tortoise, J., and Hare, W. (2007) High-velocity *versus* low-velocity strategies. In: *Studies of Quadripedal Locomotion* (F. Aesop ed.), Mythical University Press, Erewhon CA, pp. 123 - 134.
>
> Tortoise, J and W Hare. High-velocity *versus* low-velocity strategies. pp. 123 - 134, in "Studies of Quadripedal Locomotion" (ed. by F. Aesop), Mythical University Press, Erewhon CA (2007).
>
> Tortoise, J and W Hare (2007). High-velocity *versus* low-velocity strategies, in Studies of Quadripedal Locomotion (F. Aesop, ed.), Mythical University Press, Erewhon CA, pp 123–134

The easiest way to be sure that you are using the correct format for your target journal is to use the references in papers published in that journal as a template. As in the case of references to papers, it is a good idea to get the format right the first time; you will save yourself considerable time and effort later in the publication process if you proceed with care at this stage.

11.3 References to electronic sources

Electronic sources of information are proliferating even more rapidly than scientific journals and it is likely that you will have to include references to websites in your list of references and, also, in the text of your manuscript. The best way to insure that you are correctly incorporating the address (also known as the URL, which stands for uniform resource locator) of a website into your list of references or into your manuscript is to find a similar example of a web address in your target journal and to copy the format. If you are unable to find an example that seems appropriate and if the Instructions to Authors do not include specific instructions about electronic sources, you can refer to the website of the American Psychological Association (http://www.apastyle.org/elecref.html/) for useful advice and information about references to "Electronic Media and URLs." As noted on this website, you should check, at every stage of preparation of your manuscript, the address of any website to which you refer. Websites have a nasty habit of disappearing or changing their location on the internet.

When you refer to a website, be sure that the address that you give corresponds exactly to the web page that the reader of your paper will want to view. A web address that, for example, takes the reader to the homepage of a scientist is not much use if the reader is looking for very specific information that is buried somewhere in an unidentified link that is one of many on the scientist's homepage.

Finally, you should try, whenever possible, to avoid references to websites that require readers to pay for the privilege of viewing them.

11.4 Words in foreign languages

Some of the papers to which you refer may have been published in a language other than English, in a journal whose title is not in English, or, in the case of books, in a city whose name is not given in English. When you encounter such a reference, remember that the purpose of your list of references is to convey information and to allow the reader to find your sources in the literature. Thus, you should make sure that, as far as possible, all the information in your references is in English. If the title of a paper is in French, for example, you should translate the title into English and add, in parentheses, "in French." The name of the journal should, however, remain in French, in the appropriately abbreviated form, to facilitate retrieval of the paper by a reader who wants to read it in the original French. If an English abstract is included with the original paper, you should replace "in French" by "in French with English abstract." In the case of the reference to a chapter in an Italian book, for example, you should provide the title of the chapter in English with "in Italian" in parentheses. The format should be the same as that for chapters of books written in English; the name of the book should be given in the original Italian; and the name of the publisher should be in Italian for ease of retrieval but the location of the publisher should be in English, for example, Venice or Rome and not Venezia or Roma. These latter considerations also apply to references to entire books, when no chapter is mentioned.

11.5 References to papers "in press" and to unpublished data

Your target journal may provide instructions for references to papers that have been accepted for publication but have not yet appeared. In general, citation of such papers follows the standard format for references in your target journal but the volume and page numbers are replaced by the words "in press," often in parentheses. Your target journal may or may not require that you include photocopies (or electronic files) of such papers when you submit your paper.

You should not include references to unpublished results in the References section. When you refer to unpublished results in the text of your paper, you should follow your reference in the text with the words "unpublished results" and the names of the people who obtained the results, with the entire phrase in parentheses as follows: (unpublished results, B. Snow and A. White).

The same rules apply to "personal communications," namely, to results of another researcher who has not yet published them but has told you about them. In addition to following any allusion to such results in the text by the phrase "personal communication" in parentheses, you would be wise to obtain written permission from the researcher to mention his results in your paper. Unless you have such permission, you cannot be sure that he will be thrilled to see the first mention of his important results in your paper.

11.6 Notes

The reports in some journals, most notably *Science,* are followed by a section that includes both references and notes, which are intermingled and numbered sequentially. The notes provide additional information about both the background of the study and the experiments performed as part of the study. If you are hoping to publish your paper in *Science*, for example, you should study similar papers in *Science* to determine exactly what should be in your main text and what should be included as notes. Only work of general interest to scientists in a variety of fields is published in *Science.* Moreover, the editors of *Science* reject 90 percent of the papers that they receive. Thus, you should consider carefully your chances of getting your paper accepted before you take the trouble to prepare it according to the appropriate format. If you send your paper to *Science* and it is rejected, you will have to take the notes from your Notes and References section and insert them appropriately into your text to conform to the style of your second choice of target journal.

11.7 A note about the number of references

While it is essential that you cite papers that describe the experiments and discoveries that form the basis for your work and that provide any information necessary to repeat your experiments, you should try to keep the List of References as short as possible. As noted in Chapter 5:

> [In the text, it may be possible to limit the number of references by including] some version of the all-encompassing phrase, "Brainy *et al.* (2007) and references therein." Using this format, you are able to refer the reader to all the references cited by Brainy *et al.* in 2007 and, thus, to all the relevant work published before Brainy *et al.* wrote their paper in 2007.

Try to limit your references to papers that have appeared within the past few years, adding older ones only if really necessary.

Chapter 12

Figures and Figure Legends

12.1 General advice

Figures enhance the written descriptions of the results of your experiments, Nonetheless, you should also make sure that, as far as possible, all your Figures are self-explanatory. Readers may scan your paper and look at your Figures before they invest the time required to read the text of your paper. Furthermore, the material that you display in your Figures should be so clear and so convincing that the reader can easily draw the same conclusions from each Figure as you yourself have done. I have already mentioned the presentation of Figures in Sections 8.4 and 8.5 and you should refer back to these sections before proceeding further.

12.2 Graphs and histograms

Figures that show graphs should contain sufficient information to justify the space that they occupy in your paper. Conversely, they should not contain so much information that your results are indecipherable and your point is lost. For example, it is generally better to describe in words, in the main text of your paper, any relationship that can be illustrated graphically as a straight line through three or four points. At the other end of the spectrum, a Figure that includes six different curves, each of which passes through eight different points, might be too complicated for the reader to absorb easily and it might be better to include two Figures, each showing three curves. When your paper is published, the material in each of your Figures should be easily discernible without the aid of a magnifying glass.

Any symbols that you use in a Figure should be defined in the Figure Legends (also known as the "Legends to Figures" and "Figure captions"; check your journal for the correct nomenclature) but, if at all possible, you should indicate the meaning of each symbol on the Figure itself so that the reader does not have to keep referring to the legend to find out what each symbol represents. If you use abbreviations in your Figure, they should be the same as the corresponding abbreviations in your text and should have some mnemonic value, for example, "+act" for "plus actinomycin" and "+suc" for "plus sucrose."

The widespread use of computers and color printers has led to the proliferation of Figures in all colors of the rainbow when, for the most part, black and white, with cross-hatching and shading, are all that is required. Scientific journals serve a very different purpose from fashion magazines and you should restrict your use of color to those occasions when it is absolutely necessary. You can certainly prepare multicolored diagrams for oral presentations of your work at meetings and symposia but try to avoid the unnecessary use of color in Figures that will be published on paper in order to reduce the cost of reproduction.

The inappropriate use of three-dimensional histograms to depict the relationship between two variables is a bad habit that developed before the general availability of color printers. Three-dimensional histograms that represent relationships between two variables are meaningless. A histogram is also known as a bar graph. For the relationship between two variables, only two-dimensional bars are necessary. A single vertical line, extending above and below the upper limit of each bar, should show the standard deviation or standard error of the mean of the results represented by the bar.

Bars that represent the results of different measurements can be differentiated by shading of different types, such as stippling and cross-hatching. However, you should not shade bars unless it is absolutely necessary and you should avoid solid black bars whenever possible; there is rarely a good reason for using all that black ink.

12.3 Units and axes

When you draw a graph or a histogram, you must label each axis appropriately. It is also essential that you use the same units of measurements for the variables that you are displaying as you use in the text of your paper. Thus, for example, if you have given concentrations in mg/ml and rates in mol/s in the text, you should not label your axes $mg.ml^{-1}$ and $mol.s^{-1}$. Furthermore, in order not to waste space, you should not extend the axes very far beyond the extent of the largest values of the variables that you are plotting. Similarly, when you are drawing bar graphs, you should not make the bars unnecessarily wide. In other words, use space efficiently and make an effort not to cover the page unnecessarily.

12.4 Logarithmic and semilogarithmic scales

If you are plotting your data on a logarithmic or semilogarithmic scale, remember that $1 = 10^0$, that $10 = 10^1$, and that $100 = 10^2$, so that $\log_{10}1 = 0$, $\log_{10}10 = 1$ and $\log_{10}100 = 2$. Thus, on a logarithmic scale (to the base 10) that corresponds, for example, to the number of cells, values of 0, 1, and 2 correspond to one cell, 10 cells, and 100 cells and not to 0, 10, and 100 cells. Remember that zero on a logarithmic scale does not equal zero on an arithmetic scale. Zero on a logarithmic scale (to the base 10) is equal to one.

12.5 Photographs

Photographs can be very useful if the features that are important to your results are clearly visible. In many cases, black and white photographs and micrographs are adequate and color adds nothing. As a general rule, you should only use color photography when black and white photography fails to make your point. If you have a particularly dramatic color photograph that illustrates your results, you can submit it as a possible cover photograph if your target journal has an illustrated cover.

You must make sure that every photograph includes an indication of scale. If you are including a photograph of a large object or large animal, for example a horse, the scale should be obvious from the surroundings but, even in this case, a scale is useful. For photographs of small animals, plants and cultured cells in Petri dishes, for example, you can include a ruler in the photograph to provide a scale. For smaller objects and micrographs of all kinds, a scale bar on the photograph is the most effective way of indicating the dimensions of your material. It is much less helpful to include an indication of the scale in the legend to the Figure, for example, "original magnification, 100×" or "100× magnification." By contrast to such phrases, a scale bar on the photograph itself provides an immediate and obvious frame of reference.

If you use arrows and arrowheads on your photographs to draw the reader's attention to particular features, make sure that they are clearly visible and of reasonable size. You should also consider the size of letters when you label features on a photograph with abbreviations. The letters should be legible without a magnifying glass and you must explain all annotations on photographs in the legends that accompany them.

If your photograph shows bands of macromolecules that have been fractionated by electrophoresis, you should consider showing the mobility of at least one standard macromolecule for reference.

12.6 Diagrams and schemes

If you are planning to include a diagram or schematic representation (and I shall refer here exclusively to diagrams, for simplicity's sake) that summarizes your methodology or results, consider carefully how the reader will approach such material. Many authors present diagrams in an attempt to summarize methods or to simplify the interpretation of complicated results. However, such diagrams often contain so many elements that the reader becomes completely confused.

When you discuss a diagram during an oral presentation, you are able to point to each item in the diagram and lead your listeners from one item to the next in a logical sequence. By contrast, when a reader sees a diagram on a page, she sees some or all of the following items simultaneously: words; numbers; symbols; formulae; straight arrows; curved arrows; two-headed arrows; minidiagrams; and sketches of equipment. Far from helping the reader understand your ideas, the diagram gives her an instant headache. If you want the reader to understand your diagram, try to make sure that she knows where to begin, which direction to go, and where to finish. If your diagram illustrates a system with many feedback loops, it might be helpful to break the system down into its various components and present these as individual diagrams, rather than cramming all the information into a single diagram. Often, the only person who can make head or tail of the diagrammatic representation of a set of related processes is the person who drew the diagram in the first place. Therefore, you might well consider whether, rather than a picture being worth a thousand words, a thousand words might be of much greater value to your readers than a single picture.

12.7 Capitalization in Figures and diagrams

In all Figures and diagrams, single words and the first word of each short phrase should be capitalized unless they are common abbreviations that start with a small letter, such as tRNA, cDNA or cAMP.

12.8 Figure Legends

The ideal combination of Figure and legend allows a reader to understand the results in the Figure without reference to the text of the paper. Nevertheless, you should avoid very long legends in which you repeat material that is described at length in the text. In general, the first sentence or title of your legend should state the result that is depicted in the Figure. The middle part of the legend should provide the reader with some idea of how the data in the Figure were obtained. If the Figure displays quantitative data, the last sentences should explain any symbols or abbreviations in the Figure and the statistical significance of any numerical differences that are apparent in the Figure. If the Figure shows non-quantitative results or observations, you should summarize the methods used in the preparation of the material in the Figure, for example, the methods used to visualize samples after electrophoretic fractionation or to stain materials in photomicrographs. You should then explain any abbreviations in the Figure and, finally, you should include the scale of any photographs or micrographs. If, as I recommended above, you included a scale bar in your photographs or micrographs, you should indicate the length represented by the scale bar. These suggestions will not apply to the legend to every Figure, of course, but they are widely applicable.

If the summary of the relevant details of your methodology is too long to include in a Figure legend, you can refer to such details in the text of your paper by including one of the following phrases: ". . . as described in Materials and Methods," ". . . as described in Experimental Procedures," or ". . . as described in the text." When you explain your symbols, you need to remember to use the same units as in your Figure and in the main text of your paper. It is easier for the reader to understand a Figure if the symbols are explained within the Figure itself. Thus, you should try to include the definitions of symbols within each Figure, provided that you can do so without cluttering up the Figure and obscuring the results themselves.

12.9 Reproduction of Figures and Tables that have been published elsewhere

If you use a Figure or Table that has already been published, irrespective of whether it appeared in a publication by you or some other researcher, you should obtain written permission for both electronic and printed reproduction of the material from the senior author of the paper in which the Figure or the Table appeared originally (if you are not the senior author) and from the publisher of the journal in which it appeared. At the end of the legend to any previously published Figure or Table that you include in your paper, you should add the phrase, "Published with permission of . . ." and include the name of the journal.

Chapter 13

Tables

13.1 General considerations

A clearly formulated Table can provide an easily assimilable summary of a large amount of data. Tables can be part of your Materials and Methods. For example, in medical papers, a Table can be useful for summarizing the clinical details of the patients in a study and the outcome of treatment. In papers in molecular biology, Tables are useful when many lines of cells, plasmids, and sequences are discussed in the Materials and Methods. Tables are useful as part of your Results when you are discussing the values of several related parameters that do not lend themselves easily to graphical representation. On rare occasions, Tables may even help you to explain your results in the Discussion. You should not, however, put your data in a Table if they can be described simply in a sentence or two. Tables that are not strictly necessary are a waste of space.

13.2 Titles of Tables and footnotes

It is likely that your target journal will provide clearly defined parameters for the layout of your Tables and you should pay close attention to them. The title of each Table should be entirely self-explanatory but should not be longer than a single sentence. Footnotes to the title are acceptable unless specifically vetoed by your target journal. Often such footnotes refer to conditions under which measurements were made or to the units of measurement. If you do not explain the units of the various parameters in footnotes to the title, you should remember to include them, when necessary, at the tops of columns and on each horizontal line of the Table. As in your Figures, the units in your Tables should be identical to those in your text.

The items in Tables are frequently numerical and inexperienced authors tend to exaggerate the accuracy of their results by using too many significant figures and decimal places, in particular, when they cite average values or coefficients of correlation. As I noted in a discussion of this issue in Section 8.6, it is wrong to cite average values and standard deviations in such a way that the accuracy of each average and standard deviation appears greater than the accuracy of each individual measurement. Parameters that provide an indication of the significance of differences or confidence levels should also not be given to more significant figures than are necessary to demonstrate your point.

13.3 Keep it simple

The function of a Table is to provide the reader with ready access to numerous pieces of information. If such access is hindered by the presence of too much data and if the size of the print in your Table is too small, the Table will not serve its purpose. In such cases, reconsider whether it is necessary to include all the data and, if necessary, distribute your data among several Tables.

Chapter 14

Supplementary Information

Your paper should include all the results that are necessary to convince the reader that your conclusions are valid. However, you might have some results, for example, Figures and Tables, that are not essential to your argument but that strengthen its foundations. Space limitations might prevent you from including these results in your paper but many journals now allow authors to publish such results on the internet under the heading "Supplementary Information." Moreover, some editors and reviewers now suggest that authors should remove certain results from their papers and publish them on the internet as Supplementary Information. The instructions for the preparation of Supplementary Information are generally very specific and you should follow them to the letter.

You should bear in mind that Supplementary Information will be reviewed with your paper and, if you submit your paper as "hard copy," you should provide printed copies of your Supplementary Information. In general, you should not rely on Supplementary Information for support of the results in your paper. It is better to pare down your results to those that are essential to your conclusions. However, if the Supplementary Information is a short video sequence, for example, you obviously cannot include it in your paper and the internet is the only way you can make your video immediately accessible to many scientists.

Chapter 15

First letter to the editor of your target journal

15.1 The purpose of the letter to the editor

You have written your paper and now you are ready to submit it to your target journal. You have followed all the Instructions to Authors and your paper conforms exactly to all the specifications for submission by regular mail or as electronic files. The next step on the pathway that leads to the publication of your paper requires that you convince the editor of your target journal to submit your paper to the process that is known as "peer review."

The term "peer review" suggests that your paper will be reviewed by your peers. However, the definition of "your peers" is very fluid. Thus, while your true peers might be defined as scientists of your own age, who are working in the same field as you are at institutions of a similar caliber to yours, it is likely that the reviewers of your paper will be members of the journal's editorial board, scientists whose names have been suggested by members of the editorial board, scientists whose reviews the editor solicits on a regular or irregular basis, or scientists whose names you, yourself, have suggested, as noted below in Section 15.3. These reviewers are unlikely to be your peers in all the respects noted above but this discrepancy is not important and, in any case, there is nothing you can do about it. All you need to be concerned about at this stage is whether or not the editor decides to send your paper out for review or returns it to you unread.

The editor of a journal that receives many submissions is not very different from a shopper in a supermarket. As the shopper walks down the aisles, he glances very briefly at all the items on display, picks up a few items to examine them more closely and, finally, drops one or two into his basket. If a product fails to catch his eye, there is no chance that he will purchase it; if something about the item displeases him on closer examination, there is also no chance that he will purchase it. Only when an item meets all his specifications does that item find its way into his basket. You want to make sure that there is nothing about the submission of your paper to the editor of your target journal that prevents her from "buying" it.

15.2 Presentation and salutation

If the cover letter that accompanies your paper is poorly presented, the editor will be prejudiced against your paper before she even starts to read your letter. Thus, your letter should be carefully printed and laid out on the page. Make an effort to find out if the editor is a man or a woman and begin your letter "Dear Sir" or "Dear Madam." An editor does not like to be addressed "Dear Sir or Madam." If you know the name of the editor, you can write "Dear Professor Jones" or "Dear Dr. Jones," for example, but you should not write "Dear Professor W. Jones" or "Dear Prof. Jones." If your letter is well presented and your salutation shows that you have taken the trouble to find out something about the addressee, it will make a good first impression.

15.3 The body of the cover letter

After the salutation, you need to begin as illustrated by the following example:

> Please find enclosed four copies of a manuscript entitled, "Lungulin stimulates the secretion of sputase from murine HT37 fibroblasts," by Peter Brainy, Mary Gifted, and Wilbur Right.

(If you are sending your material electronically, you should replace "enclosed" by "attached.")

This introductory sentence should be followed by one sentence that describes the state of the field, one sentence that describes what you did in your study, one sentence that describes what you found, and one sentence that explains the importance of your results. A new paragraph should then begin with a sentence that explains why your paper should be of interest to the readers of the journal to which you are submitting it. The next sentences should state that all the authors contributed to the work in the paper, that they all take responsibility for it, and that none of the work described in the paper has been published elsewhere.

Some journals allow authors to suggest the names of reviewers. If you provide such names, you should make sure to include complete mailing addresses, e-mail addresses and telephone and fax numbers, if they are available. No editor is going to make any kind of effort to track down a reviewer that you suggest if you do not give her sufficient contact information. You can include this information in the body of your letter, provided that it is neatly formatted, or you can allude to it at this point in your letter and refer the editor to a list of possible reviewers that appears on a separate page.

If you have a competitor in your field who you think might review your work unfavorably or might take advantage of your unpublished results, you can state here that you would prefer that this person not review your work.

You should finish your letter by thanking the editor for considering your manuscript for publication and telling her that you look forward to hearing from her at her earliest convenience.

It used to be considered polite to close letters to strangers with the words "Yours faithfully," while letters to acquaintances ended with the words "Yours sincerely." This custom has been abandoned and "Yours sincerely" has replaced "Yours faithfully," which now appears rather stilted.

15.4 A sample letter to the editor

Here is a fictitious example that you can use to guide you in the preparation of your letter to the editor. You should bear in mind, however, that your letter must always conform to any specifications indicated in the Instructions to Authors of your target journal. This sample is meant only to serve as a guide. The letter itself would, of course, be printed on the author's official stationery or under a letter-head with the author's address and contact information. Furthermore, if the author of this fictitious letter were writing from outside the USA, he would have to add "USA" to Professor Brown's address.

Professor J. Brown
Editor, *Journal of Respiratory Sciences*
Dept. of Pathology
University of Podunk Medical School
2343 Maple Street
Erewhon, CA 12345

October 7, 2007

Dear Professor Brown,

Please find enclosed four copies of a manuscript entitled, "Lungulin stimulates the secretion of sputase from murine HT37 fibroblasts," by Peter Brainy, Mary Gifted, and Wilbur Right. Recent studies have indicated that various polypeptides alter the morphology and metabolism of murine fibroblasts. We examined the proteins that are secreted into the medium when murine fibroblasts are stimulated with lungulin, a peptide of 26 kDa that we isolated from *Hortensia multiflora*, which is a herb that has traditionally been used for the treatment of respiratory distress in Central Asia. We found that lungulin stimulated the secretion of several peptides from murine HT37 fibroblasts and identified one of the peptides as sputase by Western blotting analysis. Our results show that lungulin might be a useful drug for the treatment of respiratory diseases and suggest a mechanism for its action.

Our work should be of particular interest to the readers of the *Journal of Respiratory Sciences.* All the authors contributed to the work described in this paper and all take responsibility for it. Moreover, none of the work described in this paper has been published elsewhere.

We would like to suggest four researchers as possible reviewers of our paper and their names, with full contact information, are attached on a separate sheet. We would prefer that Dr. T. Racer of the University of Walthamstead not review our paper. Her laboratory is in direct competition with ours.

Thank you for considering our paper for publication in the *Journal of Respiratory Sciences.* We look forward to hearing from you at your earliest convenience.

Yours sincerely,

Wilbur Right Ph.D.

You may have trouble condensing your work into just a few sentences but it is essential, if you are to keep the editor's attention, that you present your work, its relevance and its importance as succinctly as possible. Sometimes it helps to think in terms of the spoken rather than the written word. If you were to meet the editor at a scientific meeting and were to ask her to consider publishing a paper that you had just finished, she would ask you what your paper was about and you would probably be able to tell her in one sentence. If she then asked you why you performed your study, what you found and why it was important, you would probably be able to answer each of these questions in a single sentence. These are the sentences that you need to include in your letter.

If your brief letter succeeds in attracting the editor's attention to your work, she can always look for more information in the Abstract or Summary of your paper. Your single goal in writing your letter is to hold the editor's interest for the time it takes her to decide between moving your paper to the next stage in the review process and sending it back to you with an apologetic but impersonal letter of rejection. A long and complicated exposition of the background to your study, copious details about your experiments and overoptimistic statements about the novelty, importance and potential impact of your results are unlikely to impress the editor more than a concise letter of the type shown above.

It is worth devoting a certain amount of time to the letter to the editor. If she does not like your letter, she will not consider your paper for publication and will not send it out for review.

15.5 Additional documents

You should check the Instructions to Authors of your target journal to determine whether you have to provide any additional documents with your letter to the editor. Such documents might include, for example, a statement signed by all the authors of your paper, in which they attest to the accuracy of the results and take responsibility for them, or a completed copyright assignment form. Failure to include the necessary documentation will probably result in automatic rejection of your paper by the editor's secretary before your paper even reaches the editor's desk.

Chapter 16

Submission of your paper

16.1 On paper, as electronic files or via the internet?

Submission of papers as electronic files (on computer diskettes or CDs) and via the internet has become routine and, indeed, some journals that are published exclusively on the internet (e-journals) only accept electronic submissions. Journals that are published in the traditional format on paper are able to minimize production costs by accepting manuscripts that are submitted as fully formatted electronic files and/or camera-ready text and Figures. In each case, the journal's publisher saves money by having you do the work that used to be done by the journal's editors, production assistants, and typesetters. If you fail to produce an electronic file or camera-ready material that conforms exactly to the specific requirements of your target journal, your paper may be rejected out of hand or publication may be significantly delayed. Before you decide how to submit your paper, if you have a choice between traditional and electronic formats, read the Instructions to Authors very carefully to make sure that you can produce the type of material that is required and can meet all the necessary criteria.

16.2 Submission on paper

If you have followed the instructions carefully as you prepared your text, list of references, Figures, and Tables, your paper should be ready for submission as hard copy, on paper. You should now check that you have used margins of the required width, the required spacing between lines and the font or fonts of the required size. You should also check to make sure that your pages are numbered appropriately.

Some journals require that the title page be page 1; others require that pagination start with the page on which the Abstract or the Introduction is printed; and some journals require a cover page, before the actual title page, with just the title and contact information of the author who is submitting the paper. You should also check to determine whether the sections of your paper are in the correct order (for example, Figure Legends before Tables or *vice versa*). When you have made sure that your text is complete and properly organized, you should check the Instructions to Authors to determine whether you need to include the number of words, the number of Figures, the number of Tables, and the words "Date submitted" and "Date accepted" at the beginning or end of your manuscript.

The Figures themselves require special attention at this stage. Should each be labeled on the front or the back? Should there be an arrow on each Figure to indicate its correct orientation? How many copies of original Figures are required? How many photocopies of each Figure are required? You will find the answers to each of these questions in the Instructions to Authors and should meet all the specific requirements of your target journal.

When you have made sure that you have met all the requirements of your target journal, you need to make the requisite number of copies of your entire paper. You should send the copies of your entire paper, plus your cover letter, to the editor of your target journal by an express service, if possible. The express mail services of national postal services and commercial carriers, such as Federal Express, United Parcel Service, and DHL, provide both speed and reliability. Moreover, express mail services allow you to track your manuscript as it makes its journey from your desk to the editor's desk and to obtain proof of its delivery as soon as it has been delivered.

Even though your manuscript is of considerable value to you, it has very limited monetary value. Thus, there is no point sending it by registered mail unless mail services between your location and the editor's are notoriously unreliable. Moreover, a packet sent by registered mail may not reach its destination for one or several weeks. Be sure to record the tracking number of your packet and the internet address of the express mail service so that you can follow the progress of your packet on the internet. Instead of waiting nervously for an acknowledgment of the receipt of your paper from the editorial office, you will know within a few days that your paper has reached its destination safely.

16.3 Submission on a diskette or a CD

The requirements for the submission of papers as electronic files on diskettes or CDs are very specific and vary from journal to journal. You must be able to meet these requirements exactly if you plan to submit your paper in this way. Here are some examples of the type of instructions for such submissions:

> Diskettes, preferably 3.5 inch (1.4 MB), should be in PC/DOS or Apple Macintosh format. The data should be saved in a pure text format (ASCII), as well as in the format of the word-processing program that was used to prepare the manuscript.
>
> Figures should be saved as separate electronic files and should not be embedded in the text. Figures that are not unusually complicated, such as line drawings, histograms, and regular graphs, prepared in Excel, should be saved as Excel files (.xls). Line drawings and graphs that have not been prepared in Excel should be scanned on a flatbed scanner.
>
> Files should be saved as TIF files. However, JPG, GIF, EPS, and BMP files are also acceptable. Figures created with software programs that use proprietary formats are unacceptable. The minimum resolution for line drawings and charts is 1,000 dots per inch.

When you submit your paper on a diskette or CD, you should label the diskette or CD with your name, the date, the title of your paper, the names of the individual electronic files and the name of the software program that you used to create each file, and the type of computer that you used. You will, of course, have to send a cover letter, on paper, with your diskette or CD and, in most cases, you will also have to send a printed copy of your paper. Thus, in addition to preparing your electronic files correctly, you will need to pay attention the instructions for the preparation of hard copy that are given in the preceding section. You should send the entire packet by express mail service, as also discussed in Section 16.2.

16.4 Electronic submission

The fastest way to send your paper to the editorial office of your target journal is via the internet. If you are not familiar with the production of computer files of text and illustrations and if you have a choice between electronic submission and submission on paper or on a CD or diskette, you should hesitate before choosing this route. You should also bear in mind that your time is better spent thinking about and doing experiments than trying to generate perfectly formatted computer files. Unless you know what you are doing, you should ask someone with experience in the electronic transmission of papers to help you. If you do not know what you are doing and nobody is available to help you, you should, if possible, submit your manuscript on paper instead of wasting time and getting frustrated at your computer.

The instructions for electronic submissions are generally very detailed and specific, as illustrated by the following example:

> The journal will only accept the text of your paper as a Microsoft Word file created with MS Word 6.0 or a later version. Do not embed Figures in the text. We will convert MS Word, TIFF, and EPS files into PDF files for you. All illustrations must be provided as TIFF or EPS files. A list of acceptable Mac OS and Microsoft Windows graphics applications can be found at http://cpc.cadmus.com/da/applications.asp. For graphics, we cannot accept application programs such as Microsoft Office, Corel Perfect Office, Lotus Smart Suite and SigmaPlot.

If this example seems like gibberish to you, you should try to avoid submitting your paper as an electronic file. If you want to familiarize yourself with the skills needed for electronic submissions, choose a time when you are not anxiously trying to submit your most recent research for publication. Now is not the time to figure out the difference between a TIFF file and an EPS file or to learn how to use a new application program. If you can submit your paper electronically as easily as falling off a log, there is no better way to do so. A perfectly prepared electronic submission, once it has been reviewed and accepted, will sail through the production process and appear without delay.

Chapter 17

Letter from the editor and your response

If you did not receive your paper back by return of post because you failed to meet some obvious criterion for submissions to your target journal, you can expect to wait several weeks or even months before you hear from the editor. However, your manuscript might be returned almost immediately if you sent your paper to a journal, such as *Nature* or *Science,* that rejects 90 percent of submissions, most often without sending them out for review.

When you do finally hear from the editor of your target journal, she will tell you that your paper has been accepted without revisions, that it has been accepted with revisions, that it has been rejected but she will welcome resubmission of your paper when you have made some major changes or improvements, or that your paper has been rejected entirely.

17.1 Acceptance without revision

Editors very rarely accept scientific papers for publication without any revision. If your paper is accepted outright, you have good cause for celebration. Take your co-authors out for tea, a beer, or a dinner with champagne!

17.2 Acceptance with revisions

If your research is of good quality and you have met all the criteria for submissions to your target journal, it is likely that your paper will be accepted conditionally, that is to say, with revisions. The fact that your paper has been accepted with revisions means that the editor has read the reviews of your paper and decided that the changes that the reviewers have requested are reasonable and that you will be able to follow their suggestions without difficulty and improve your paper accordingly. Thus, acceptance with revisions is also cause for modest celebration, provided that the celebrants remember that they still have a certain amount of work to do. Revisions after a paper has been accepted conditionally do not usually involve many new experiments. Generally, the reviewers ask for more information about experiments that you have already performed and for a more detailed Introduction or Discussion or, perhaps, a reanalysis of your data by a specific method.

17.3 Rejection with an offer to reconsider

If you receive a letter of rejection that includes an offer by the editor to reconsider your paper after you have performed certain additional experiments or revised the text and Figures extensively, your course of action is clear. Do not consider sending your paper, unchanged, to another journal. You will be shortchanging yourself and your collaborators if you ignore the reviewers' advice, which is likely to help you improve your paper considerably. Go back to the laboratory and set up the additional experiments or sit down at your desk and revise your manuscript as recommended by the reviewers.

When you have met all the reviewers' suggestions, resubmit your paper with a new cover letter, based on the example in Section 15.4, that includes sentences modeled on the following template:

> In your letter of December 10, 2007, you offered to reconsider the possible publication of our paper, entitled, "The expression of receptors for lungulin on murine fibroblasts," after we had performed the additional experiments suggested by the reviewers. We have completed all the suggested experiments and enclose four copies of a revised version of our manuscript.

These sentences should be followed by sentences that describe the background to your experiments, your experiments, your results and their implications, as indicated in Section 15.4. Your final sentences should read as follows:

> We thank you for the opportunity to resubmit our manuscript to the *Journal of Respiratory Science* and hope that it is now suitable for publication. We look forward to hearing from you at your earliest convenience.

17.4 Outright rejection

Do not despair if your paper has been rejected outright. Several options remain open to you. If the journal to which you sent your paper has a very high rate of rejection of manuscripts and your manuscript was rejected without even being sent out for review, you need only make few changes to your paper before you resubmit it to another journal. The changes that you will have to make will be exclusively those that are necessary to insure that your revised manuscript conforms to the formatting requirements of your new choice of target journal. Once you have made these changes, you can immediately and optimistically resubmit your manuscript to this journal. If your manuscript was sent out for review and then rejected, you should study the letter of rejection and the reviewers' comments carefully and use them to improve your manuscript and its chances of publication. If the editor or the reviewers pointed out a major flaw in your experiments or in your reasoning, you now have the opportunity to correct your mistakes and produce a publishable piece of work. No matter whether you have to perform more experiments and/or rewrite your manuscript, you should eventually be able to resubmit your work in an acceptable form to a journal that will be happy to publish it.

There is, of course, a hierarchy of journals in every field. The journals at the top of the heap tend to be those that publish papers of the broadest general interest and, thus, they have the largest circulation and the greatest number of readers. As journals become more and more specialized, their readership decreases and a decrease in readership is considered to reflect a decrease in prestige. A journal dealing with a very narrow field and having a relatively small circulation might be considered less prestigious than another journal with a larger circulation. However, the papers in the former journal might well be of the same quality as those in the latter journal. If your paper has been rejected by a journal that caters to a large audience, consider sending it to a journal that is targeted to specialists in your field. Inevitably, this journal will have a smaller readership but, among those people who do read the journal, the proportion that is likely to appreciate your paper will be larger.

Chapter 18

Second letter to the editor with responses to reviewers

18.1 Your second chance

Your second letter to the editor, in which you respond to the suggestions and criticisms of the reviewers and, perhaps, to comments made by the editor herself, is your second chance to get your paper accepted. It might also be your last chance to get your paper accepted by this particular journal. Therefore, your second letter and the accompanying responses to reviewers must be a model of clarity and must address every issue that was raised by the editor and reviewers.

It is possible that the reviewers who commented on your original manuscript will also read the new version of your paper. However, you cannot be sure that such will be the case and, in any case, you should assume that, at best, the editor and the reviewers will have only the vaguest recollection of your original paper. Therefore, you should compose your letter to the editor and your responses to reviewers such that anybody should easily be able to understand exactly what changes you have made and exactly why you made them. To make your task easier, imagine, as you write this letter, that the editor with whom you dealt has moved to another position and that the reviewers who will consider the revised draft of your paper will be different from the original reviewers.

18.2 The second letter to the editor

You should begin your second letter to the editor by citing the title of your original paper and the names of the authors. You should address your letter to the editor to whom you addressed your first letter unless you know that she has, in fact, moved to a new position. If you know that the journal has a new editor, you should address the new editor by name, if possible. If the editor assigned an identification number to your paper, you should include this number, as follows:

> Dear Professor Brown,
>
> re: "Lungulin stimulates the secretion of sputase from murine HT37 fibroblasts" by Peter Brainy, Mary Gifted, and Wilbur Right; ms. no. 123–03.

You should start the text of your letter with an expression of polite gratitude, such as, "We are most grateful to you and the reviewers for the helpful comments on the original version of our manuscript. We have taken all these comments into account and submit, herewith, four copies of a revised version of our paper."

If the only criticisms and suggestions for improving your work came from the editor, you should address them in your letter at this point. However, it is likely that the comments by the reviewers required fairly detailed responses, which you have written out separately. Thus, the next part of your letter will include some version of the following two possibilities:

(i) In response to comments from the editor:

> In your letter of December 10, 2007, you suggested that we should replace Figure 6 by a Table. We have done so and the data in Figure 6 are now presented as Table 3. You also suggested that we examine the statistical significance of the results in Figure 7 (in the original version). We used Student's *t*-test to examine the significance of differences, as indicated in the revised Materials and Methods, and the results are indicated by asterisks in Figure 6 (in the revised version), with an explanation in the legend to this Figure.

(ii) In response to the comments from reviewers:

> We have addressed all the comments by reviewers A and B, as indicated on the attached pages, and we hope that the explanations and revisions of our work are satisfactory.

The details given in (i), the first example, provide all the information that the editor needs to retrieve her first letter to you and to confirm immediately that you have addressed the issues that she raised. However, if another editor were to read this letter, she would also be able to tell immediately how you had addressed the requests made by the first editor and what these requests were. You should prepare your responses to the reviewers similarly, as discussed in the next section.

In the final sentence of your second letter to the editor, you should return to polite formalities, as follows:

> We hope that the revised version of our paper is now suitable for publication in the *Journal of Respiratory Sciences* and we look forward to hearing from you at your earliest convenience.
>
> Yours sincerely,
>
> Wilbur Right Ph.D.

18.3 Responses to reviewers

If you were lucky, the reviewers of your paper listed their comments in numerical order. If these comments were written as continuous text, you would be wise to convert the text into a set of numbered comments before you start revising your paper. You should not ignore any critical comment made by any of the reviewers and you should address each comment separately, even if comments by one reviewer are identical to those by another. You should also group all the responses to each reviewer separately.

It is appropriate to begin each set of responses to reviewers with a few polite introductory sentences, for example:

> We are grateful to reviewer A for the critical comments and useful suggestions that have helped us to improve our paper considerably. As indicated in the responses that follow, we have taken all these comments and suggestions into account in the revised version of our paper.

You should then address each of reviewer A's comments, either in the numerical order in which they were made or in the numerical order into which you divided them. Each comment and response should be able to stand alone, without reference to the original communication from the reviewer. The reviewer's comments should be quoted *verbatim* (word for word) if possible or, if necessary, with minimal paraphrasing for grammatical or syntactical reasons. Each of your responses should be an entire grammatical sentence. No abbreviations should be included unless they are defined in the responses to each reviewer. Any references cited should be included in full, with titles. Here is an example that you can use as a template.

> Comments by reviewer A.
>
> Comment #1.
> The authors should replace Figure 6 by a Table.
> Response.
> In the revised version of our paper, Table 3 includes the data from Figure 6 of the original paper and Figure 6 in the original paper has been removed.
>
> Comment #2.
> The authors should examine the statistical significance of the results in Figure 7.
> Response.
> We used Student's *t*-test to examine the significance of differences among the results shown in Figure 7 in the original paper, as indicated in the revised Materials and Methods. The results of this analysis are indicated by asterisks in Figure 6 in the revised version (removal of the original Figure 6 necessitated the renumbering of the Figures), with an explanation in the legend to Figure 6.
>
> Comment #3.
> The authors fail to acknowledge the contributions of Helfstein's group in the Discussion.
> Response.
> We agree that we should have mentioned the recent work by Helfstein's group and we have included a reference to their work (J. Horstein and T. Helfstein, 2007, The role of lungulin receptors in health and disease, in Studies in Pneumonia and Bronchitis, vol. 27, pp. 347–354, Verdana Press, Parkville NJ) in the Discussion on page 23 of the revised manuscript.

When you have addressed each of the comments by reviewer A, you should repeat the process for reviewer B, starting with the same polite introduction:

> We are grateful to reviewer B for the critical comments and useful suggestions that have helped us to improve our paper considerably. As indicated in the responses that follow, we have taken all these comments and suggestions into account in the revised version of our paper.

Then you should address each of the comments made by reviewer B, even if some of the comments are the same as those made by reviewer A. Thus, if the editor sends the revised version of your paper back to the original reviewers, each reviewer will receive a personalized set of responses to the comments that he or she made. Alternatively, if the editor chooses to determine herself whether you have satisfied the reviewers and responded appropriately to their suggestions, she will be able to see at a glance how you have addressed the individual comments made by each reviewer.

When you have completed your revisions of your paper and your responses to the reviewers, you should check to determine whether you need to recount and restate the number of words, Figures and Tables in your paper. You should also double-check to make sure that you have referred to the correct pagination (page numbers) in the new and the old versions of your paper in your responses to the reviewers. If the numbering of Figures and Tables is different in the revised paper from what it was in the original version, you should recheck that, when you refer to the Figures and Tables by number in your revised text, you are referring to the correct Figures and Tables. Similarly, if your references are in numerical order and you have inserted some new references, make sure that all the references in the text and in your revised list of references are numbered correctly.

If you are sending your manuscript as hard copy, be sure to make the appropriate number of copies of the revised paper and Figures and send them to the editor with your new cover letter and the comments for reviewers. You should send the entire packet back to the editor in the same format as you sent your original manuscript (hard copy, diskette, or CD by express mail service; or electronic files via the internet).

Chapter 19

Congratulations, your paper has been accepted!

Congratulations. The editor has approved the revised version of your paper and has accepted it for publication. If it is your first independent paper, you should be very excited. If your paper is to appear exclusively on the internet in an "e-journal," you might not have long to wait before you see it on the World Wide Web.

If your paper was accepted by a journal that is published on paper, you will probably have to wait several months until your work appears in print. Moreover, after you have completed a few necessary formalities, for example, the assignation of copyright to the journal, and have requested a certain number of reprints (offprints) over and above those that you will receive at no charge, it is likely that you will not hear directly from the journal until your paper is returned to you as galley proofs or page proofs. When you receive these proofs, you will also receive instructions on how to correct any errors that have appeared during the prepublication process. When you correct your proofs, you should not introduce any new material into your paper. If it is necessary to add some details to correct a mistake in your results or conclusions, you can do so but you should bear in mind that you will be charged for making any substantive changes. If you need to make changes of this type, you should contact the editorial office to discuss the changes that you want to make and to determine how much you will be charged for making them.

If you submitted your paper electronically, you might not see your paper again until it is published since the publisher of the journal will use your electronic files, just as you sent them, without any changes.

Some publishers provide a web-based system, whereby an author can monitor the progress of his manuscript after it has been accepted for publication. Such systems are becoming more common and allow many authors to follow their papers through the publication process without having to communicate directly with the editor or the editorial office.

19.1 Prepublication publicity

Even if you think that you might have made an extraordinarily important contribution to your field and would like to tell the world, acceptance of your paper does not mean that you should immediately call a press conference to announce your results. The publishers and editors of peer-reviewed journals take a very dim view of scientists who publicize their results in the local or national media prior to publication. If you think that your work deserves publicity, you should contact the editor of the journal in which your paper will be published and discuss the possibility of a press release on the day that your paper will be published. You might also contact the press officer of your institution, who will probably be able to advise you on how best to draw attention to your research and, by extension, to the institution itself.

While you might be impatient to publicize your work, you can take comfort from the fact that the date of acceptance of your manuscript does establish your historical claim to the discoveries in your paper. You are free to include your data in oral presentations and can now do so knowing that nobody can take credit for your work by repeating it and submitting it for publication.

The best reaction to the acceptance of your paper for publication is to return to the laboratory and continue the research that will eventually lead to your next publication. Good luck.

Appendix

A note about writing applications for financial support

Each organization that provides funding for research has its own specifications for grant applications and these specifications can be quite complicated. Nonetheless, you will find that much of the advice in the main body of this book is as relevant to writing a grant application as it is to writing a paper. Thus, if you are about to write a grant application, you should reread Chapters 2 through 8 at least.

It is easier to write a paper that is accepted for publication than it is to write a successful application for funding. Moreover, it is an unfortunate truth that the people who read a grant application often do not put as much effort into reading it as the applicant deserves. Therefore, the easier it is for reviewers to read a grant application and to find their way through the maze of sections, headings, and subheadings, the more likely it is that the applicant will be able to convince them that the proposed research should be funded. There are two straightforward rules that can facilitate both the writing and the subsequent reading of a grant application: (i) follow the instructions and (ii) organize all sections of the application consistently.

The first rule is self-evident and you should spend enough time reading the instructions to become totally familiar with them. You should also refer to them continually as you prepare the various sections of your application. Most funding agencies require you to state what you want to do, what you have done already, how you are going to do what you want to do, and what you are going to do with the results that you get. The formal versions of these requirements generally fall under headings such as "Specific Aims," "Preliminary Results," "Experimental Details," and "Conclusions." I shall refer to these headings exclusively but the points that I emphasize should be relevant to any application for funding, no matter what its exact format.

There is a very simple way to organize your grant in such a way that the reviewer never gets lost or frustrated while reading it. After a brief introduction, you should list each of your specific aims with numbered headings and, if necessary, numbered subheadings. Then, if you listed, for example, five specific aims, you should also divide your Preliminary Results into five sections with the corresponding numbers. In the first section of the Preliminary Results, you should discuss your first specific aim and the progress that you have made towards achieving this aim. If you described your first specific aim using numbered subheadings, you should use the same subheadings and numbering in your Preliminary Results. Thus, anyone reading your application who might wonder how much progress you have made towards achieving each or any particular one of your specific aims will be able, immediately, to locate the relevant material in your Preliminary Results. Thus, for example, if the aim in question was the second of your specific aims, the preliminary results related to this second aim should be listed in the second section of Preliminary Results.

You should apply the same numbering system to your Experimental Details. All the experimental details related to your first specific aim should be in the first numbered section of the Experimental Details. If your first specific aim was listed with a series of subheadings, you should provide the details of your proposed experiments with the same subheadings in the same order. By writing your grant in this way, you will make it easy for the reviewer to follow your train of thought and your plans. If the reviewer is particularly skeptical about your fourth specific aim, she can turn first to the fourth section of your Preliminary Results and then to the fourth section of your Experimental Details, with each section being immediately recognizable both by its number and by its heading.

Your Conclusions should be organized in the same way as the preceding sections so that there is a clearly defined path from your specific aims, via your preliminary results and the details of your experiments, to the possible conclusions to be drawn from your proposed research.

If you follow these suggestions, you will find not only that you have written a grant application that is easily navigable by a reviewer but also that you have clarified your own thoughts and plans. Furthermore, if you have to write a Progress Report during the period for which you receive funding, you will find that the task will be very simple because all you will need to do is to describe the progress that you have made towards each of your specific aims, just as you listed them in the original application for funding. The numbering and headings in your Progress Report should correspond to those in your original application. If you have made some unexpected discoveries that do not fall under your original headings, you should discuss them after you have discussed your progress towards each of your original specific aims.

Valedictory

All the advice in this little guidebook can be summarized as follows: always read and follow the instructions; and always strive for clarity and consistency.